The Global SILICON VALLEY Handbook

The Official Entrepreneur's Guide to the Hottest Startup Scenes from around the Globe

MICHAEL MOE

and the Global Silicon Valley Team

GSV.com

GRAND CENTRAL PUBLISHING

NEW YORK BOSTON

Grand Central Publishing
Hachette Book Group
1290 Avenue of the Americas, New York, NY 10104
grandcentralpublishing.com
twitter.com/grandcentralpub

First Edition: March 2017

Grand Central Publishing is a division of Hachette Book Group, Inc. The Grand Central Publishing name and logo is a trademark of Hachette Book Group, Inc.

Print book interior design by Timothy Shaner, NightandDayDesign.biz

Library of Congress Cataloging-in-Publication Data

Names: Moe, Michael, author.

Title: The global Silicon Valley handbook : the official entrepreneur's guide to the hottest startup scenes from around the globe / Michael Moe and the Global Silicon Valley Team.

Description: First edition. | New York, NY : Grand Central Publishing, [2016] | Includes bibliographical references.

Identifiers: LCCN 2016025223| ISBN 9781455570324 (trade pbk.) | ISBN 9781478968382 (audio download) | ISBN 9781455570317 (ebook)

Subjects: LCSH: New business enterprises—Finance. | Venture capital. | Entrepreneurship. | Enterprise zones.

Classification: LCC HG4027.6 .M64 2016 | DDC 658.1/1—dc23 LC record available at https://lccn.loc.gov/2016025223

ISBNs: 978-1-4555-7032-4 (trade paperback), 978-1-4555-7031-7 (ebook)

Printed in the United States of America

Q-MA

10 9 8 7 6 5 4 3 2 1

ACCLAIM FOR The Global SILICON VALLEY Handbook

Dina

"It had to happen! Silicon Valley is now ubiquitous! *The Global Silicon Valley Handbook* gives you facts and an informed opinion about everything you need to know when starting your company."

—Dick Kramlich, chairman and co-founder, New Enterprise Associates

"Entrepreneurs do change the world. We are entering an exciting new era where you can go from an idea to reaching millions of people at breathtaking speeds. *The Global Silicon Valley Handbook* is chock-full of invaluable information on who you need to know, what you need to know, and where you need to go."

—Carol Bartz, former CEO of Autodesk and Yahoo!

"Michael Moe has long been an influential thought leader at the cutting edge of global technology, investing, and leadership. In *The Global Silicon Valley Handbook*, Michael has compiled an invaluable, clever, insightful, incredibly practical, and fun resource for entrepreneurs, VCs, and anyone interested in the innovation economy in Silicon Valley and around the world."

—James M. Citrin, leader, Spencer Stuart CEO of Practice, and author of *The Career Playbook*

"Very enjoyable. A great tongue-in-cheek guide for entrepreneurs on where to be seen, who to raise capital from, and how to pitch. It'll even tell you which restaurants to frequent and which phrases never to say to a VC. Michael Moe has created an unfiltered insider's playbook to navigating the Global Silicon Valley. I'll be consulting it next time I'm abroad."

—Joe Lonsdale, founding partner, 8VC

"Finally! A book from a true Silicon Valley ecosystem innovator! *The Global Silicon Valley Handbook* is an entertaining 'must read' for anyone interested in how this Valley works. Michael Moe takes you around the Valley with an entertaining tongue-in-cheek edge, while giving you the smarts you need to become a success in the tech world. Dig in. You will read it again and again."

—Tim Draper, founder, Draper Associates and DFJ

"Michael Moe and the Global Silicon Valley Team have written the playbook on how an entrepreneur can win big in the Global Silicon Valley.

—Ronnie Lott, NFL Hall of Fame and VC investor

"Finally, the definitive source of what all entrepreneurs really need to know! From Silicon Valley to Singapore, this book tells you whom you should be meeting with, where to hold those meetings, what to say, what questions to ask, what books to read, and so much more. This is a true Handbook full of useful info. Seriously, there is no B.S. If you want to start a company you need to have this at your side."

—Dave Pottruck, former CEO, Charles Schwab, and Active Angel Investor

To all the entrepreneurs, who know there is nothing more powerful than a big idea and have the guts to make it happen.

To my colleagues at GSV, who believe in the dream and the magic of growth.

To my friend and coach Bill Campbell, who showed how caring—more than people thought possible—could change the world for good.

And to my family, especially my wife, Bonnie, and my daughters, Maggie and Caroline, who amaze and inspire me every day.

Contents

PART 1 INTRODUCTION

OTHER NEED TO KNOWS

PART 2 NAVIGATING THE TOP 50 INNOVATION CENTERS

PART 1

INTRODUCTION

ABOUT THIS MANUAL

AMERICA WAS BUILT ON THE BELIEF THAT EVERY PERSON IS GUARANTEED THE RIGHT TO LIFE, LIBERTY, AND THE FOUNDING OF A STARTUP. Yet somehow conventional wisdom has created a laundry list of requirements to be an entrepreneur: Drop out of Harvard, sit in on a calligraphy class to develop multiple typefaces and proportionally spaced fonts, build a computer in a garage, etc., etc. This kind of thinking is *wrong!* In fact, we contend that there is only one requirement to being a successful entrepreneur: reading this handbook.

The Global Silicon Valley Handbook aspires to have the answers to every question a budding entrepreneur might have, with features on every major startup city in the world, from Seattle to Sydney to Stockholm to Seoul. It'll tell you which restaurant in Shanghai to close the deal in, what words never to use in a meeting with Andreessen Horowitz, which filters to use on Snapchat, and so much more.

The Global Silicon Valley Handbook aspires to have the answers to every question a budding entrepreneur might have, with features on every major startup city in the world.

You'll be able to go from pitch to public offering with ease, knowing which accelerator to trust and which coffee shop to build your business in, when to zip your hoodie and when to pull out your BlackBerry. (*Hint: Never.*) And you'll be raising more and more venture capital money every single step of the way.

Stop thinking in terms of which college you graduated from, who your father is, what company you interned at, and when you started coding. Entrepreneurs aren't all rich, white, male, twenty-something-year-old Libertarians. They're people who have recognized problems and are building solutions—**PEOPLE LIKE YOU.**

THE USER'S GUIDE TO THE GLOBAL SILICON VALLEY HANDBOOK

While being an entrepreneur can be amazingly fun and inspiring, it's also hard. In order to help, GSV's goal is to distill the most interesting tidbits of information meant to help you build the next great startup. Inside the book, we include the following sections to help you go from being an immigrant to becoming a native in the Global Silicon Valley.

KEY EVENTS TO BE AT

We list the key events to be at. Why? Two key things beyond your own talent that you need to be a successful entrepreneur are access to other talent and access to money. Being where the "right people" hang can make it smoother sailing to get your idea off the ground. Also, having fun is an important ingredient for long-term success.

VENTURE ACTIVITY

This provides data from online startup and VC databases on how active the specific market is for attracting venture capital investment. While it's true that money will find a great idea anywhere, if you're trying to catch fish, you want to go where the fish are. By understanding and following the venture activity, you can find the right pond to land in.

TOP ANGELS

Angel investors are usually different from venture capitalists, typically taking less time and doing less due diligence before they write a check. Hence, knowing the right angels in your region is important to get your idea going. However, angel investments are only a piece in the puzzle, as more institutional capital will be required to fuel the company beyond the early stages.

TOP ACCELERATORS AND INCUBATORS

Replacing the garage and your best friend's spare office, incubators and accelerators have emerged to support the early days of a startup's development. The best incubators and accelerators provide not only office space, but also insights, resources, and moolah.

BEST BARS, COFFEE SHOPS, AND MORE

See "Key Events To Be At"...but let's face it, more relationships have been created over a beer than a business plan meeting, so we list the best of the best out-of-office settings to do business. Additionally, VCs and angels are people, too, and are often more approachable...and interesting...after a few cold ones.

GUIDE ON HOW TO DRESS

Dressing for success has been an important business fundamental since Lancelot wore armor in battle. But what that means depends on the geography and the environment. Standard Global Silicon Valley attire is emerging for entrepreneurs: with the cool T-shirt, funky tennis shoes, and badass attitude being main staples.

IN THE KNOW

We've included timeless advice and insights from locals who built the Global Silicon Valley. While what becomes a big idea will be different, the components of what makes a big idea are unchanged. Follow these tips to live like a local and jumpstart your journey to greatness.

HOTEL

LETTER TO THE READER
CEO MICHAEL MOE

Welcome to the Global Silicon Valley!

In 1849, pioneers from around the world came to California looking for gold, and what once was an empty landscape of natural beauty turned into a densely populated melting pot. A hundred and sixty something years later, the gold is long gone but the land of opportunity continues to BOOM. People still flock to California to chase a dream, but now one that is intermixed with microchips, apps, social networks, and most important: Silicon Valley.

At the heart of the tech revolution still in mid-swing, Silicon Valley is home to the superstars Apple, Google, and Facebook; the old-timers Intel and HP; and of course, the thousands of little guys still trying to make it: the startups. But Silicon Valley has quickly become more than a geographic spot on a map; it is a mind-set of innovation that has spread across the globe, and that's what we at GSV stand for.

We're delighted to share our vision of uniting a community of innovators who all hold a passion for changing the world for good.

One hundred years ago, Silicon Valley was filled with apricot and apple orchards with a twenty-year-old fledgling university called "The Farm." Detroit was the innovation capital of the world with the young and ambitious migrating to the Motor City to seek fame and fortune. Automobiles were the disruptive technology, and Henry Ford was the Steve Jobs of his day.

Marked by prominent bold letters, the Pioneer Building houses GSV Asset Management. As the first building seen on Woodside Road west of I-280 toward the mountains, it embodies GSV's expectation to pioneer the industry in finding, investing in, and partnering with the "Stars of Tomorrow."

Today, Detroit is bankrupt, and Silicon Valley has been the Mecca for entrepreneurs from around the world who believe that there is nothing more powerful than a big idea. Stanford University is Startup U, where people flock to pursue their dreams. Over the past fifty years, it's been breathtaking to see how companies conceived on a napkin at Buck's and started in garages between San Francisco and San Jose have transformed society and business.

Apple, Google, Facebook, Oracle, Cisco, eBay, Intuit, Intel, Tesla, Uber, Twitter, Lyft, Coursera, and Dropbox are but a few of the revolutionary companies that call Silicon Valley home.

While the sixty miles between San Francisco and San Jose remain the epicenter for innovation, the spirit and entrepreneurial mind-set that made Silicon Valley such a magical place have gone viral and global. From Austin to Boston, to Chicago and São Paulo; from Shanghai to Mumbai to Dubai, there is something very powerful happening: *a Global Silicon Valley is emerging.* And it doesn't have to rhyme with anything, either.

The Global Silicon Valley Handbook aspires to inspire the next Steve Jobs or Elon Musk to create the future they can imagine.

GSV created a formula for determining the top fifty markets in the Global Silicon Valley based on factors such as VC funding, startup activity, educational institutions, and business climate. We then researched each market to highlight key features and asked the local experts to understand what makes their city tick. We answer entrepreneur need-to-knows on how to be successful in bringing their concept to life. Who are the top angels? Which are the top incubators? What should be in the pitch? What shouldn't be in the pitch? Where is the power breakfast spot? What is the dress code? And much, much more…

The Global Silicon Valley Handbook aspires to inspire the next Steve Jobs or Elon Musk to create the future they can imagine. We are believers in how one person's vision can change the world and that whatever the mind can conceive and believe, it can achieve.

That's what this handbook is about and why it is the official guide to the Global Silicon Valley.

Read on!

—MM

GETTING IN THE DOOR

Before you launch the next great startup, take a step back and look at the big picture. Over the past 15 years, venture capital funding and initial public offerings have accelerated at an astounding rate. Just look at the evidence.

VC + IPO ACTIVITY: 2000 vs. 2015

INDICATOR	2000	2015
Number of Unicorns	1	133
NASDAQ 100 P/E	105x	22x
IPOs	445	152
IPO Proceeds	$108 billion	$25 billion
Median Time to IPO	3 years	10 years
Median IPO Offering Size	$84 million	$92 million
Average IPO First-Day Pop / Year-End Gain	53%/–19%	16%/–0.4%
Percentage of Profitable IPO Companies	26%	30%
Median Annual Revenue of IPO Companies	$18 million	$39 Million
VC Investments	$105 billion	$59 billion
Late Stage VC Investments	$18 billion	$16 billion

Source: National Venture Capital Association, PwC, WilmerHale, *Fortune*

The dramatic increase in the number of large private companies is mainly a result of two key trends.

First, venture capitalist (VC)–backed private companies are staying private longer (a median of three years in 2000 versus ten years in 2015). Second, there are now over 3 billion people on the Internet with 2.6 billion smartphones, enabling entrepreneurs to go from an idea to a product that reaches billions of people at warp speed.

WORDS TO KNOW

INDICATOR

(in·di·ca·tor) *noun*

These are the series of metrics and statistics we consider to be the most relevant toward understanding the differences in the private and public markets.

We are riding powerful tailwinds that are rapidly transforming the world as we know it. In 2000, there were only 370 million people on the Internet (roughly 6% of the world population), no one had heard of a smartphone yet, broadband was a fantasy, and mobile applications off a platform had not been invented. Today, the "digital tracks" have

BUBBLE vs. BOOM

INDICATOR	2000	2015
Internet Penetration	370 million (6%)	3.1 billion (43%)
Broadband Penetration	60 million (1%)	2.3 billion (32%)
PC Penetration	180 million (3%)	1.4 billion (20%)
Mobile Phone Penetration	740 million (12%)	7 billion (98%)
Smartphone Penetration	0	2.6 billion (28%)
Tablet Penetration	0	500 million (7%)
Mobile App Downloads	0	226 billion
Computing Cost (per person)	$7.03	$0.04
Computer Storage Cost (per person)	$4.77	$0.02
Digital Natives in Workforce	6%	35%
Global Middle Class	1.4 billion	2.5 billion

Source: Gartner, Nielsen, A. T. Kearney, eMarketer, KPCB, GSV Asset Management

been laid and 140 billion apps have been downloaded from Apple and Google.

Moreover, in the Internet Bubble of 2000, the ten largest Internet companies were valued off a figment of one's imagination. Now, they are mainly valued on future cash flows discounted back to today. Apple, with a $640 billion market value, has $200 billion of sales and $190 billion in cash.

However, as those who have seen Benchmark Capital's VC Bill Gurley's tweetstorm know, valuations for private companies are reaching sky-high levels. On the one hand, the value of every Unicorn put together is less than the market capitalization of Facebook, and lower valuations means less VC money being spent. On the other hand, many Unicorns are reaching huge valuations with little profit to show for it, meaning a market correction may be in order. As Gurley argues, this means there could be a shift in venture capital from focusing on growth to focusing on the path to profitability.

WORDS TO KNOW

TWEETSTORM

(tw·ee·t·st·orm) *noun*

A series of tweets, usually 5 or 6 in a row, that are posted back to back to back on Twitter. Users post tweetstorms about a topic they want to voice their opinion about... but can't summarize in 140 characters

TIP

When somebody asks you a direct question, give a direct answer. Telling the investor that you'll get back to them later is a sure way to have the investor tell you that they'll get back to you later.

HOW VENTURE CAPITAL WORKS

Venture capitalists (VCs) are people, too—most of them anyway. And as such, while insanely ambitious, competitive, and hopefully intelligent, just like anyone else, they aren't looking for more work to do.

A blind business plan hurled over the e-mail transom has as much chance of being read as an e-mail starting out with "Dear Sir or Madam." A top VC might receive 500 to 1,000 e-mails a day, so cluttering up the inbox without an edge is only making you the enemy of the assistant whose job it is to delete e-mails. You might think you have the best idea since Travis hitched a ride on Uber, but without a warm introduction from a trusted person in their network, your idea will die an anonymous death.

LinkedIn can be a start to finding out who you are "linked" to, helping you build a path toward your VC target. Unfortunately, many people are LinkedIn whores and send and accept LinkedIn invitations from people they barely know. Affinity groups are a good way to get your foot in the door. Going to the same college or high school as the targeted VC is super helpful. VCs will be courteous to somebody who comes from their hometown (they don't want to be called the rich jerk who forgot where he came from) or from their alma mater.

Don't send a business plan. That takes too much time to read and is often impossible to understand without an interpreter.

Networking through past business affiliations of the VC is one of the best ways to get a meeting. If a VC worked with someone who can attest to your brilliance or your startup's potential, you now have provided the VC the trusted shortcut to finding the *next big thing*.

Don't send a business plan. That takes too much time to read and is often impossible to understand without an interpreter. A short deck describing how your idea is going to transform the world is what will get their

> **TIP**
>
> In Silicon Valley, having business failures on your résumé is viewed positively (you learned something), but working at a bureaucratic brain-dead organization is viewed like cancer.

THE TOP

The best place to start a startup is in school . . . and you don't even have to finish! Here are the top universities that produce the most startups.

TOP 10 GLOBAL UNIVERSITIES BY VC-BACKED FOUNDERS PRODUCED (2010–2015)

Rank	School	Founders	Companies	$$ Raised
1	Stanford	378	309	$3.5B
2	UC Berkeley	336	284	$2.4B
3	MIT	300	250	$2.4B
4	Indian Institutes of Technology	264	205	$3.2B
5	Harvard University	253	229	$3.2B
6	University of Pennslyvania	244	221	$2.2B
7	Cornell University	212	190	$2.0B
8	University of Michigan	176	158	$1.2B
9	Tel Aviv University	169	141	$1.3B
10	University of Texas at Austin	150	137	$1.3B

attention. Include the 4P description of your business—who are the **people**, what is the **product**, what is the **potential**, and what is the **predictability** on execution. Once you've got the meeting, the pitch is all about precision and passion—don't speak in vague generalities or like a professor. No VC wants to understand how to build the clock—they want to know what time it is.

The beginning of the pitch basically sets the stage for success or to be sent packing. The first two minutes are about getting the VC engaged and establishing credibility. Recognizing that they reject 99.9% of the investment opportunities presented to them, you need to create the impression that what you have is revolutionary and that you are Marc Benioff's better-looking clone.

Next you need to explain the market opportunity that you are going after. What is the TAM (Total Addressable Market)? What are the megatrends that your idea benefits from? Who are the competitors?

Now it's time to explain your plan for world domination. What is your go-to market strategy? How does this disrupt the status quo? What kind of people and partners do you need to execute against your opportunity?

Finally, explain the financial model. What are the key metrics that will drive value and how are you focused on achieving success? List the key milestones that will provide confirmation that you are on the right track to take over the industry.

After going through your presentation, don't overstay your welcome. Make sure you give the impression that you are tight on time and you are on your way to your next meeting, whether you are or you aren't. Back to VCs are people, too—they want something that others want or can't have. The harder you are to get now, the more they will want you. We just solved all your dating problems, too.

GSV's 4P's

Investors aren't always articulate on what they are looking to invest in other than "making money," but actually, we believe that 99% of the characteristics investors want are captured by the 4P's—**PEOPLE**, **PRODUCT**, **POTENTIAL**, and **PREDICTABILITY**.

The first "P"—**PEOPLE**—is by far the most important component to the investment equation. While startups don't have long histories, the people do. Winners find a way to win and also know how to get you to believe in what something can become.

The second "P"—**PRODUCT**—is critical, with investors looking to back companies that can be the leader in what they do. With technology in general and the Internet in particular, there is often a disproportionate advantage to the leader in a category.

The third "P"—**POTENTIAL**—is what investors need to envision. They want to believe that the fledgling enterprise could be HUGE. Megatrends pour gas on a fire for startups.

The fourth "P"—**PREDICTABILITY**—might seem like a challenge, since there is nothing predictable about a startup other than its unpredictability. However, creating confidence that a company can deliver against its plan (see first "P) is the difference between a startup being funded or failing.

THE TOP 50 INNOVATION CENTERS OF THE WORLD

1. Silicon Valley
2. New York City
3. Silicon Beach
4. London
5. Boston/Cambridge
6. Beijing
7. Shanghai
8. Chicago
9. Berlin
10. Tel Aviv
11. Paris
12. Toronto and Waterloo
13. Austin
14. Bangalore
15. São Paulo
16. Seattle
17. Singapore
18. Amsterdam
19. Denver and Boulder
20. Baltimore–DC–VA Metro Region
21. Miami
22. Mumbai
23. Atlanta
24. Seoul
25. Stockholm
26. New Delhi and Gurgaon
27. Dallas and Fort Worth
28. Dublin
29. Dubai
30. Pittsburgh
31. Minneapolis
32. Montreal
33. Philadelphia
34. Vancouver

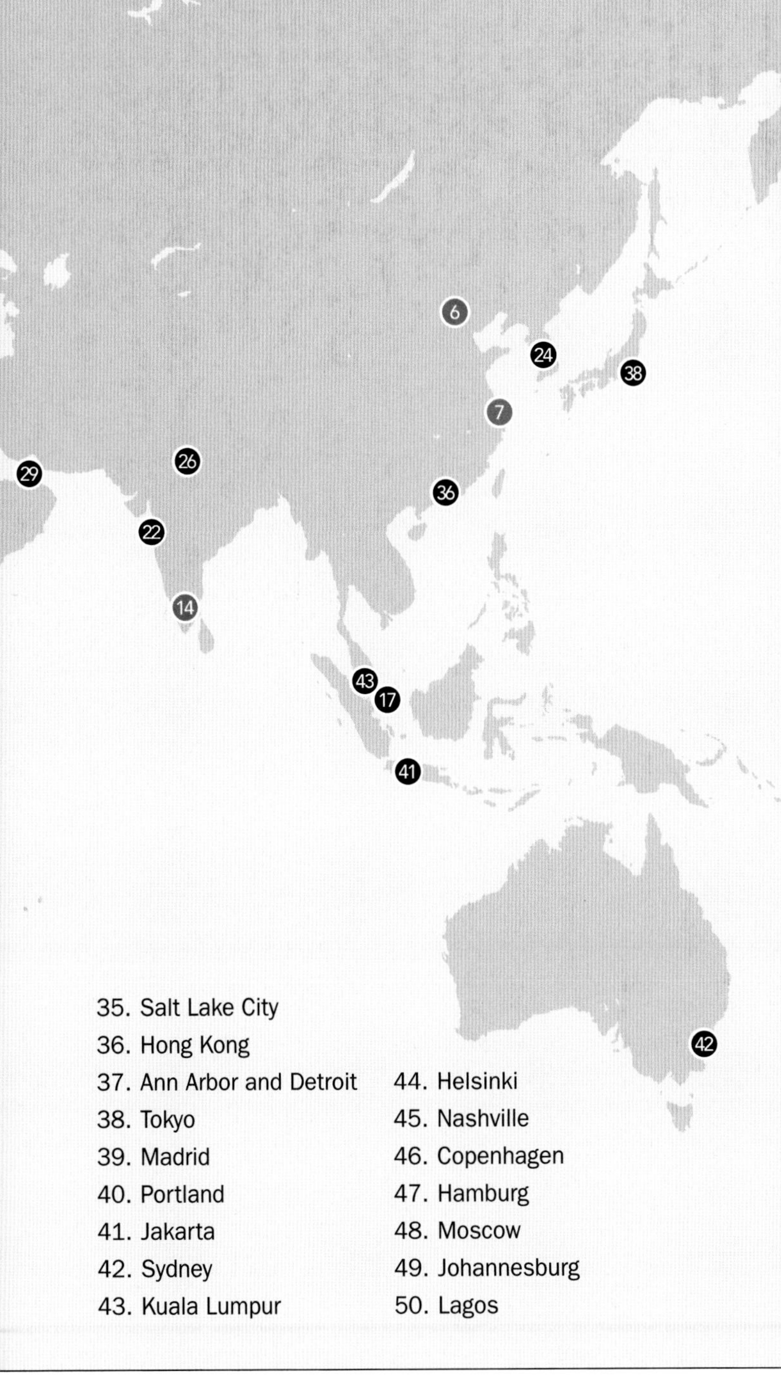

35. Salt Lake City
36. Hong Kong
37. Ann Arbor and Detroit
38. Tokyo
39. Madrid
40. Portland
41. Jakarta
42. Sydney
43. Kuala Lumpur
44. Helsinki
45. Nashville
46. Copenhagen
47. Hamburg
48. Moscow
49. Johannesburg
50. Lagos

We used our proprietary formula (i.e., it is slightly better than pulling it out of a hat) to determine the top innovation markets in the Global Silicon Valley. In the following pages, we highlight the top things to know and places to go to get your idea off the ground and into the skies.

WHAT WE LOOKED AT

- Year-over-Year Funding Growth (over 2 and 5 Years)
- Year-over-Year Deal Growth (over 2 and 5 Years)
- Venture Capital Funding in Q4'15 - Q3'16
- Number of Deals in Q4'15 - Q3'16
- Number of Exits in Q4'15 - Q3'16
- Year-over-Year Exit Growth (over 2 and 5 Years)
- Number of the Fortune Global 2000 in Each Region
- Presence of Google and Facebook Offices in Each Region
- Number of Accelerators
- Number of Top 500 Global Universities
- Average Internet Download Speed
- GSV-Conducted Survey Responses

PART 2

NAVIGATING THE TOP 50 INNOVATION CENTERS

1

SILICON VALLEY

One Tesla goes by. Then another. Then another, except this one has a bumper sticker with "My other car is autonomous but I never drive it" written on it. Then Mark Zuckerberg appears, walking with his Philz with one hand and collecting every Facebook search from his over 1.6 billion monthly active users in the other. Welcome to *the* Silicon Valley.

Of the many entrepreneurship hotspots, Silicon Valley is truly the only one that embodies a mindset and culture. Here, we detail some of the quirks that make up Silicon Valley, from its language to its fashion to its college nucleus.

AGGREGATE BAY AREA VENTURE ACTIVITY Q4'15 - Q3'16

VC FUNDING	$33.6B
DEALS	2518
EXITS	527
5 YEAR YoY FUNDING GROWTH	20.8%
5 YEAR YoY DEAL GROWTH	6.4%

Source: *CB Insights*

HOW TO DRESS

THE FOUNDER

(aka: everyone in Silicon Valley)

HOODIE
Zipped for meetings, unzipped for coding, and generally a little too loose.

CORPORATE T-SHIRT
Usually an established tech company, whether or not it is the current company of the employee, and always gotten from a college career fair.

WEARABLE TECH
A Snapchat Spectacle, Apple Watch, and arc reactor replacement heart are all necessary.

MESSENGER BAG
Worn loosely over the shoulder to show that he doesn't care if it drops; his employer will replace it.

JEANS
A little too loose or a little too tight, but a good fit is a bad idea.

SNEAKERS
The sportier the better to compensate for a lack of regular exercise.

THE INVESTMENT BANKER

Working hard to look West Coast cool, should maybe try harder.

APPLE WATCH
Signaling reverence for the King of the Valley.

HAIR
Gelled hair works well after hours at the Balboa Café but screams "frat boy" to Nerds running startups.

GLASSES
Warby Parker glasses with no prescriptions projecting Hip and Smart.

SMILE
Practices confident smile in front of mirror, saying "I'm better than Frank Quattrone" ten times before he starts his day.

SUIT
Bosses in NYC enforce a strict suit dress code, but Valley ethos fries people with ties.

DR. SEUSS SOCKS
Carried over from his Wall Street training program days.

BRIEFCASE
Usually stuffed with Bitcoins taken in lieu of real cash with startups.

THE ENGINEER

Looks like a hobo, but has the same agent as Major League Baseball players.

LOOK
Permanently stoned look caused by being permanently stoned.

HAIR
Shares barber with Jesus, thinks that he can also walk on water.

HOODIE
Unzipped to show off his personal statement.

COLORFUL WATCH BAND
Another way to say "bite me" to the establishment.

NAME BADGE
Shows that he takes pride in being "one of the people" and consistent with the cubicle culture and isn't afraid of a little kindergarten throwback.

JEANS
Jeans signal that he's a member of the creative class.

TENNIS SHOES
Hip tennis shoes with no athletic purpose whatsoever.

THE LAWYER

Double-books time in ten-minute increments, allows getting paid in stock in the next Facebook-type opportunities.

LOOK
Permanent look that she would like nothing better than to rip your heart out and sauté it.

BLAZER
Blue blazer conceals the knife set she carries around to fillet opponents.

DIET
Has her own table at the Village Pub with strict instructions to leave the dressing on the side.

SKIRT
Short skirt to have eyes notice, stare, and a bare knee will soon follow to the groin.

HEELS
Stiletto heels used to stomp on any foes and clients that welch on their $1000-per-hour fees.

HOW TO: (TALK LIKE A) VALLEY GIRL

For a full glossary of Silicon Valley lingo, see page 211.

ACQUIHIRE Acquiring a company for its employees rather than its product, almost like dating someone exclusively for their Cara Delevingne eyebrows.

ANGEL INVESTOR An affluent individual whose kids have gone to college, leaving him or her with too much liquid money and a slot-mania addiction.

BIG DATA Only Palantir knows, deep within the vaults of Thiel's mansion next to a room of students he convinced not to go to college.

BOOTSTRAP When an entrepreneur uses personal finances to back their own company and the most cited reason for an entrepreneur's divorce.

THE VENTURE CAPITALIST

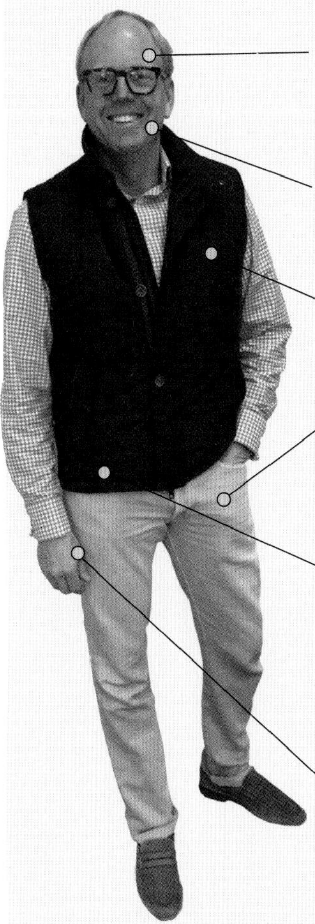

WEALTH
Forehead used to display "W2" while trolling at the Rosewood.

SMILE
Grin is highlighted by twice-a-day Crest whitening strips.

VEST
Vest is a perfect garment for nippy Valley mornings and looking like a "regular guy."

PANTS
Ditto for chinos and suede loafers, although outfit costs $1500 at Wilkes Bashford.

FIGURE
Zero body fat from twice-a-day personal trainer workouts, his own chef, and radical experimental no-fat drugs still in phase-1 trials with the FDA.

TAN
An impossible tan from many hours on the golf course and in Napa Valley Vinyards.

A SILICON VALLEY GUIDE TO SHOES

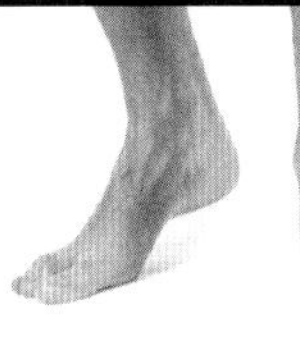

BAREFOOT
You're either a monk or a Steve Jobs wanna-be.

SANDALS
Showing confident coolness (and perhaps cluelessness) when pitching to VCs. Socks optional, but not recommended. Pedicures are not just for the ladies, gentlemen.

LOAFERS
A VC's footwear of choice, made from suede from baby cows with exotic Italian yarn stitching.

SPINNING SHOES
Often seen on those limping from their 6 a.m. SoulCycle class.

SNEAKERS
Worn by engineers solely for comfort. These shoes have never seen the inside of a gym before.

BYOD Bring Your Own Device, often to work, aka watching livestreams of the Warriors games without worrying about all the data-collecting, productivity-inducing, outsourcing-threatening mechanisms that could be in a work computer.

CLOSE Signing all the legal papers and documents to finalize an investment. See Wikipedia for "It's not over until the fat lady sings."

WORK-LIFE BALANCE Legends have been told about this mysterious concept, but the millions of twenty- to thirty-year-old college graduates in the Valley consider it a myth.

MVP Minimumly Viable Product. A product with just enough features to show early adopters its potential, providing a feedback loop to guide future improvements.

"MUST BE AT" **LOCAL EVENTS**

TECHCRUNCH CRUNCHIES FEBRUARY

Called the "Academy Awards" of technology and Silicon Valley, the Crunchies award and celebrate the top venture capitalists, entrepreneurs, angel investors, corporate innovators, and more in their various industries. There are so many awards given out that everyone walks away with something, like peewee soccer. It's Silicon Valley patting itself on its back, giving it an "attaboy." It's of course more dynamic when T. J. Miller is the MC roasting TK's lady.

BAY TO BREAKERS MAY

What do you get when you combine Halloween, a marathon, and rampant underage drinking? Welcome to Bay to Breakers, an annual footrace through San Francisco. Bring your costumes and your Redbull, as you'll be attempting to complete a 12k while drinking more booze than water.

APPLE WWDC JUNE

The perfect event to realize that your iPhone is obsolete and all the cool kids are getting new ones, so you have to, too. The World Wide Developers Conference is also famous for setting the leading fashion trends in Silicon Valley, the turtleneck-jean combo no longer "in."

ESCAPE FROM ALCATRAZ TRIATHALON JUNE

If you're not quite right in the head, check out the Escape from Alcatraz Triathlon, which includes a 1.5-mile swim, an 18-mile bike ride, and an 8-mile run. Most would rather try the actual prison escape.

THE TOP

ANGEL INVESTORS

Chris Sacca
Dan Rosensweig
Dave McClure
Dave Pottruck
David Sacks
Harris Barton
Laurene Powell Jobs
Marc Benioff
Marissa Mayer
Mark Zuckerberg
Mike Maples
Mitch Kapor
Nikesh Arora
Ron Conway
Ronnie Lott
Sam Altman

VC GENERAL PARTNERS

Aileen Lee
Ben Horowitz
Bill Gurley, Mitch Lasky + the Benchmark Boys
Bryan Schreier
David Sze
Fred Harman
Ira Ehrenpreis
Jay Hoag
Jim Breyer
Joe Lonsdale
Jon Callaghan
John Doerr
Marc Andreessen
Michael Abbott
Michael Moritz
Peter Bell
Peter Thiel
Ping Li
Sandy Miller, Todd Chaffee + the IVP team
Scott Sandell
Tim Haley
Vinod Khosla

OUTSIDE LANDS MUSIC & ARTS FESTIVAL AUGUST

The most popular music festival in the Bay Area continues to attract some of the biggest names in the music industry, and is a must go-to with friends. Just keep track of them! A Coachella for the nerdier set.

BLUE ANGELS OCTOBER

Though not supporting Christopher Columbus's legacy of slavery and European diseases, the United States Navy's pilots put on a breathtaking display at San Francisco's Fleet Week on Columbus Day weekend. No chance of scurvy, just connections.

SALESFORCE DREAMFORCE OCTOBER

One of the largest tech conferences in the World, Dreamforce attracts over 170,000 attendees annually. The conference is so big that it completely shuts down parts of San Francisco. While some people go to learn about innovation, many others go just to see what shoes Marc Benioff will be wearing.

GSV PIONEER SUMMIT SEPTEMBER

Some pitches from the ABC show *Shark Tank*: an alarm clock that wakes you up to the smell of bacon, a fart-smelling candle, and the "Cougar Lifestyle Shot" energy drink. With ideas like these, it's easy to forget that, around the world, there are brilliant innovators working day and night to do something meaningful. This summit is helping you remember, featuring a collection of leading innovators working on challenging problems whom we call the GSV Pioneers.

TOP Q4'15 - Q3'16 FUNDING ROUNDS

Uber—$3.5B
Airbnb—$1B
Lyft—$1B
Palantir—$880M
Sungevity—$650M

LAWYERS

Bob Gunderson
Gordy Davidson
Ken Guernsey
Larry Sonsini
Troy Foster

BANKERS

Frank Quattrone
George Lee
Michael Grimes
Thom Weisel

NAME DROPS

Bill Campbell
Marc Andreessen
Peter Thiel

LOCAL HEROES

Bill Campbell
Bill Hewlett
Dave Packard
Elon Musk
Gordon Moore
Marc Andreessen
Marc Benioff
Mark Zuckerberg
Steve Jobs

NAVIGATING **STANFORD**

Without Stanford, there would be no Silicon Valley, and without Silicon Valley, there would be no Stanford.

THE DOs

✔ **GO TO LECTURES** You never went in college, but now that you're trying to make it in the real world, Stanford lectures are worth the visit. Go to explorecourses.stanford.edu and choose classes to attend in every subject, from strategic management at the Graduate School of Business to introductory undergraduate computer science. The best part? You don't have to sell an arm and leg to pay the sixty-thousand-dollar tuition.

✔ **GET UP IN THE CLUB** What happens when you get seven thousand overachieving undergrads in one place? Seven thousand different clubs. Whatever you'd like to accomplish at Stanford, there's likely a club for it. Entrepreneurship clubs at Stanford include BASES, SENSA, and ASES.

✔ **HAVE THE RIGHT PITCH** Whether you're trying to recruit potential employees at a career fair or potential cofounders in class, know that Stanford students are absolutely saturated with everything about entrepreneurship. When you do talk to them, remember to appeal to their values of innovation and creativity and their "I'm God's gift to earth" complex.

THE DONTs

✘ **WALK DOOR TO DOOR ON SAND HILL ROAD** As good as your idea may be, the venture capitalists on Sand Hill Road don't want to see you banging on their door with your pitch. If you can't seem to find a venture capitalist interested in your idea, *Shark Tank* it. That's a verb, right?

✘ **NETWORKING ATTEMPTS WITH CHILDREN OF FAMOUS PEOPLE** When you're looking for an internship at Microsoft, it might seem easier to become friends with Bill Gates's daughter instead of going through round after round of grueling technical interviews. However, you'll quickly find out that these founders can barely get their kids a job, let alone you.

✘ **BEG FOR DOWNLOADS** The only thing harder than making Stanford students move to the East Coast is making them download your app. Even if it holds the lexicon of the world unknown, is it really worth 10 seconds of download time and 7 megabytes of space?

PHRASES TO AVOID IN A PITCH

The goal is to start up, not start down.

"We can always hire someone for that."

That's like looking deep into Vinod Khosla's eyes and telling him heart to heart that you want to spend his money, not make him any.

"We're like the Uber for _______."

Unfortunately, there's already an Uber for everything. It's called Uber.

"It's like Facebook on steroids."

Can't wait to see your social network develop acne, infertility, and male pattern baldness.

"User acquisition is our biggest focus right now."

This summons the image of a founder going door to door, begging anyone and everyone to use his or her product because he or she has a family to feed.

"We have no competition."

Either you're delusional or you're in the wrong industry.

"We're not afraid to pivot."

You've lost faith in your idea, your cofounder has lost faith in your idea, and now we've lost faith, too.

"This is the Holy Grail."

Your company isn't valued at $1 million, let alone eternal life.

"Our additional features are really the selling point, not the product."

But if the product is lame, why would we use the extra features?

"Worst-case scenario, we're bought by Google at a one-billion-dollar valuation."

Worst-case scenario, your company is unsuccessful on delivering its outlandish promises and fails so violently the normally amicable Ben Horowitz outsources you to a silicon factory for the rest of your life.

The Six Fastest Ways To Lose Credibility

1. Use a ThinkPad.
2. Wear a tie.
3. Drink Dunkin' Donuts Coffee.
4. Be a Dodgers Fan.
5. Still have a BlackBerry.
6. Drive a truck.

THE OLD PRO
Palo Alto

A PALO ALTO classic and GSV favorite, The Old Pro's Silicon Valley crowd and mechanical bull draw people from across the Bay. If you're in the mood to watch games, drink beer towers, and guess the net worth of any one table of people, this is the bar for you.

DO'S AND DON'TS OF **NETWORKING**

In Silicon Valley, your network is your net worth. New in town and want to make a splash? Use these tips and soon you'll be worth more than Zuck, though he's 1.6 billion friends ahead of you.

- ✗ **Stuff food in your pocket**—you don't want to be mistaken for the homeless crowd on Market Street. Plus it'll make your thighs look big.
- ✗ **Pitch an investment through LinkedIn Messages**—you'll have more success getting to a VC through carrier pigeon.
- ✗ **Stalk on Snapchat**—that's just plain creepy.
- ✗ **Talk about politics or religion**—unless you want to talk about how lame it is that Barack Obama still uses a BlackBerry.
- ✗ **Hand out your business plan**—the trick is to get somebody to beg you for it.
- ✗ **Return projectile Fireball back on the bar**—you'll kill the party.

- ✓ **Talk less; smile more**—as Aaron Burr said, "Don't let them know what you're against or what you're for."
- ✓ **Use mouthwash or breath mints**—no one will want to talk to you if you still smell like the onion bagel you had for breakfast.
- ✓ **Remember a person's name**—a person's name is their favorite sound.
- ✓ **Hand over your phone, not a card**—there's no young person that hands out a card.
- ✓ **A shot of Fireball**—you'll be the life of the party.

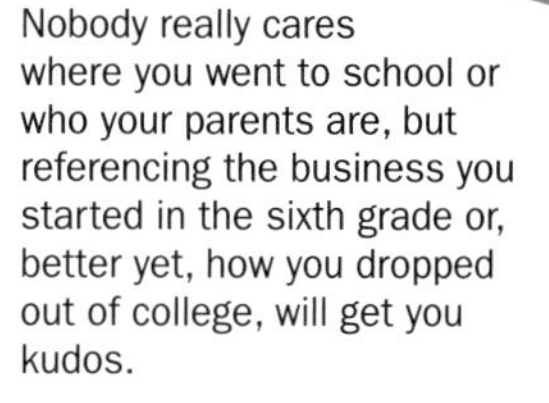

Nobody really cares where you went to school or who your parents are, but referencing the business you started in the sixth grade or, better yet, how you dropped out of college, will get you kudos.

SILICON VALLEY AT A GLANCE

Find a Silicon Valley crowd in the following places.

Power Breakfast

Buck's of Woodside, Woodside
Ann's Coffee Shop, Menlo Park
Stacks, Menlo Park
Il Fornaio, Palo Alto
Madera, Menlo Park

Clubs to Weasle an Invitation To

Menlo Circus Club, Atherton
Olympic Club, San Francisco
San Francisco Golf Club, San Francisco
Sharon Heights Golf & Country Club, Menlo Park
The Battery, San Francisco

Offline Tinder

DNA Lounge, San Francisco
The Patio, Palo Alto
The Rosewood, Menlo Park

While it's tough not to have fun at the beautiful Rosewood hotel, with its delicious drinks and dining, amazing service, and "$$$$" rating on Yelp, its famous Thursday "Cougar Night" isn't for everyone. Take a middle school dance, make these women over thirty-five and recently divorced, and make these men either over fifty and kinda creepy or under twenty-five and really into technology, and you have "Cougar Night." Cougars are definitely on the prowl, but before becoming their meat, one should conduct some thoughtful and careful soul-searching of what they want in life.

Pizza Fuel

Pizzeria Delfina, Palo Alto
Amici's, Menlo Park

Closing Dinners

Village Pub, Woodside
The Old Pro, Palo Alto
Evvia, Palo Alto
Chantilly, Redwood City
John Bentley's, Redwood City
Kokkari Estiatorio, San Francisco
Boulevard, San Francisco

Coffee Shops to Write a Business Plan In

Blue Bottle, Palo Alto
Coupa Cafe, Palo Alto
Sightglass, San Francisco
Philz Coffee, Palo Alto

Philz is the go-to coffee shop for the young and up-and-coming students, entrepreneurs, and coffee snobs. Its famous drink is the (now patented!) Mint Mojito, and though you could still probably make this coffee at home, only Philz has the ambience, youth, and finesse of making it a staple beverage.

TIP

The quicker you're able to tell your story, the greater the interest will be. Your "elevator pitch" should explain what makes your product special, how fast you can grow, and how big you can be.

THE SILICON VALLEY **PARKING GARAGE**

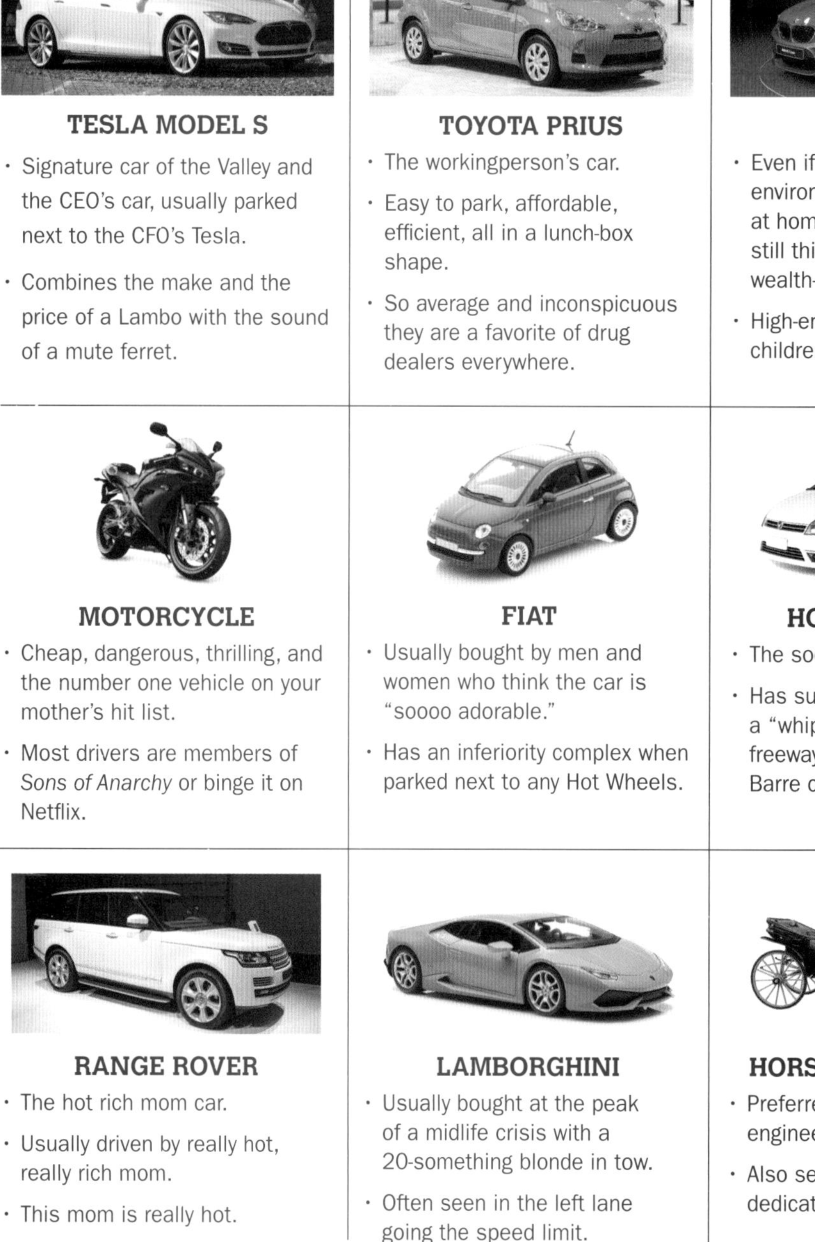

TESLA MODEL S

- Signature car of the Valley and the CEO's car, usually parked next to the CFO's Tesla.
- Combines the make and the price of a Lambo with the sound of a mute ferret.

TOYOTA PRIUS

- The workingperson's car.
- Easy to park, affordable, efficient, all in a lunch-box shape.
- So average and inconspicuous they are a favorite of drug dealers everywhere.

BMW M3

- Even if a driver is a kind environmentalist and volunteers at homeless shelters, passersby still think he is a "show-offy, wealth-obsessed d-bag."
- High-end car driven by the children of hardworking parents.

MOTORCYCLE

- Cheap, dangerous, thrilling, and the number one vehicle on your mother's hit list.
- Most drivers are members of *Sons of Anarchy* or binge it on Netflix.

FIAT

- Usually bought by men and women who think the car is "soooo adorable."
- Has an inferiority complex when parked next to any Hot Wheels.

HONDA ODYSSEY

- The soccer mom or dad car.
- Has surprising ability to be a "whip," gunning down the freeway to make it to her Pure Barre class on time.

RANGE ROVER

- The hot rich mom car.
- Usually driven by really hot, really rich mom.
- This mom is really hot.

LAMBORGHINI

- Usually bought at the peak of a midlife crisis with a 20-something blonde in tow.
- Often seen in the left lane going the speed limit.

HORSE AND CARRIAGE

- Preferred by the Amish Google engineer.
- Also seen with very, very dedicated hipsters.

You're not an entrepreneur if . . .

- You ask, "Steve who?" or "Elon who?"
- You check when the bank holidays are for your office.
- You get a speeding ticket and you don't pitch your startup to the officer.
- You don't have a TGIS (Thank God It's Sunday) because that's when your week starts.
- You hire a consultant to see if your plan is feasible.
- You start your business with the goal of selling it.
- You're at a Giants game and you don't have at least one idea for an app that would revolutionize baseball.
- You haven't maxed out at least two credit cards keeping your company afloat.
- You haven't been told by your friends to be "realistic" and "get a real job."
- Your significant other tells you "It's me or your business," and you don't say, "I love you and I'll miss you."

BAY AREA TO STANDARD AMERICAN ENGLISH TRANSLATOR

by Louis Weinstein, Published by Timothy McSweeny's

Bay Area: "Who are you again?"

English: "What company do you work for?"

■

"Do you live in this neighborhood?"

"Can you afford to live in San Francisco?"

■

"Our new mobile optimization will change the Internet as we know it."

"My startup has yet to go public and is actively seeking angel investors."

■

"What's your Facebook?"

1. "I would like to mine the Internet for information about you that basic social customs preclude me from directly asking about." 2. "I would like to stalk you."

■

"I'll take a bottle of the most expensive wine you have."

"I have enough money to get away with how insufferable I'm about to be. Yes, I am twenty-two."

"Thanks for your interest, but we've decided to promote Mitch."

"Good luck being a woman in this city. Please, for both of our sakes, don't show emotion."

■

"We should get lunch."

"I've identified your value relative to me and wish to start the process of exploiting your talents."

■

"Where are you now?"

"As our relationship is entirely based on our previous coworking experience and this social occasion has mandated that we talk, please define yourself through your current employer. This way, we can have something to discuss until we reach a mutual, satisfactory level of social interaction or I determine that your employment may be of value to me, at which point I will suggest that we 'get lunch.'"

■

"Cozy studio in south Nob Hill, $2,400/mo."

"Extra closet space in former meth lab, Tenderloin district. You will have two roommates and no kitchen. Actually $3,845/month. Partial roof."

Stargazing Atherton

Named by *Forbes* as the most expensive ZIP code in America, A-town is home to Silicon Valley's biggest players (and their ex-wives). Dozens of billionaires live in the five-square-mile area, hang out at the Menlo Circus Club, and send their kids to the elite Sacred Heart Prep School. **Welcome to the 94027.**

A. PAUL ALLEN *(Cofounder, Microsoft)*
Purchased this 22,000-square-foot estate in 2013 with 8 bedrooms, 9 bathrooms and 7 fireplaces...one for every day of the week.

B. ERIC SCHMIDT *(Chairman, Alphabet)*
Lives in a relatively modest 5,000-square-foot home. Thank you for allowing us to evade everyone's privacy... what goes around, comes around.

C. SHERYL SANDBERG *(COO, Facebook)*
The 11,500-square-foot former home of Sheryl Sandberg, with 7 bedrooms and 8.5 bathrooms, sold in 2014 for $9.25 million.

D. LARRY ELLISON *(Chairman, Oracle)*
Another one of Ellison's 30+ homes, it sold in 2015 for $25 million so Ellison can live full time in his Japanese dingy down the street.

E. JILLIAN MANUS *(Silicon Valley VC and Philantropist)*
Throws the best parties on the Peninsula.

F. MARC ANDREESSEN *(Founder, a16z)* **+**
BILL GURLEY *(General Partner, Benchmark)*
The real-life Seinfeld and Newman frenemies live a 9-iron away from each other.

G. MENLO CIRCUS CLUB
No clowns or trapezes, but Sunday Polo attracts plenty a *Pretty Woman* and flowing Napa Chardonnay.

H. CHARLES SCHWAB *(Founder, Charles Schwab)*
Online broker lives on the other side of El Camino in Atherton...not sure if he lives in a house or a forest.

I. MEG WHITMAN *(CEO, HP)*
Went from putting garage sales online to being chief of the original business that started in a garage.

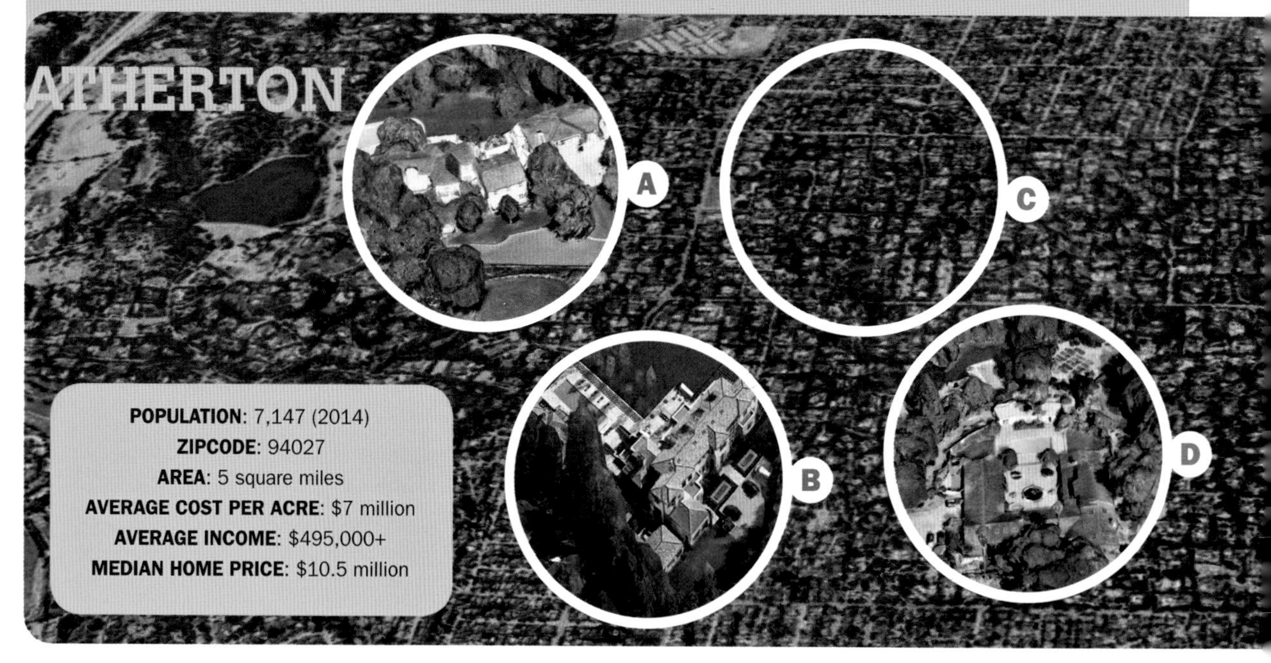

Greatest Pivots

We all remember Zuck famously pivoting Facemash into Facebook, but you'll be surprised to know that many of our favorite companies were originally duds. Match the mistake with what became magic.

Match the PIVOT. . .

1. **TWITTER** Short-form communication ☐
2. **YOUTUBE** Video + music platform ☐
3. **SLACK** Business Communication Tool ☐
4. **INSTAGRAM** Photo-sharing platform ☐
5. **NOKIA** Mobile electronics ☐
6. **SONY** Consumer electronics ☐
7. **GROUPON** Group couponing service ☐
8. **EMC** Enterprise data storage ☐
9. **YELP** Business review website ☐

. . . with the ORIGINAL IDEA

A. E-mail recommendation engine

B. "Hot or Not" dating platform

C. Rice cookers

D. Rubber boot manufacturer

E. Furniture retailer

F. Gaming website

G. Consumer tipping platform

H. Location check-in service

I. Personal podcast sharing platform

1: I; 2: B; 3: F; 4: H; 5: D; 6: C; 7: G; 8: E; 9: A

HONORABLE MENTION: Neighboring town **WOODSIDE** is also home to behemoths of its own...including Larry Ellison's 23-acre Japanese-style estate worth over $200 million. Built in 2004, his home has over 10 buildings, a man-made lake, a tea house, a koi pond and more.

LESSONS FROM SAM ALTMAN AND SILICON VALLEY

We love the Y Combinator president, and his startup series is a must watch for any entrepreneur, but watching countless lectures isn't always the most efficient. The solution? Pair episodes of "How to Start a Startup" with episodes from HBO's ***Silicon Valley***, which are hilarious, stereotypical, and informative.

ALTMAN	*SILICON VALLEY*	WHAT YOU WILL LEARN
Lecture 1: **HOW TO START A STARTUP** *(Sam Altman, Dustin Moskovitz)*	Season 1, Episode 1: "Minimum Viable Product" Season 1, Episode 8: "Optimal Tip-to-Tip Efficiency"	• The idea and the product are very different things. • Being a startup founder is incredibly hard and will suck up years of your life. • The only reason to start a startup is if there's no way you can't start it.
Lecture 2: **TEAM AND EXECUTION** *(Sam Altman)*	Season 1, Episode 2: "The Cap Table" Season 1, Episode 6: "Third Party Insourcing"	• Hiring is a last resort, and is incredibly expensive and risky. • When hiring, the absolute best people need to be hired—mediocrity will kill a company. • If people aren't working, they need to be fired fast, no matter how painful.
Lecture 3: **BEFORE THE STARTUP** *(Paul Graham)*	Season 2, Episode 3: "Bad Money" Season 2, Episode 5: "Server Space"	• Startups will consume your life. • Startups are weird, and unexpected problems with unexpected solutions will present themselves. • Expertise is helpful but by no means necessary.
Lecture 4: **BUILDING PRODUCT, TALKING TO USERS, AND GROWING** *(Adora Cheung)*	Season 2, Episode 6: "Homicide" Season 2, Episode 8: "White Hat/Black Hat"	• Interaction with users and consumers is very important for product development. • Stealth mode is stupid, and you shouldn't be paranoid about people stealing your idea. • Pivoting is an art, and not always necessary.
Lecture 5: **COMPETITION IS FOR LOSERS** *(Peter Thiel)*	Season 1, Episode 4: "Fiduciary Duties" Season 2, Episode 1: "Sand Hill Shuffle"	• The goal of a company is to build a monopoly. • Imitating others and looking at competition as validation will not get you anywhere. • Competition makes you better, but don't be blind; never stop asking yourself bigger questions.

1.5

SAN FRANCISCO

Home to hipsters, hippies, high rental rates, and even higher gentrification, San Francisco is Silicon Valley's Wes Anderson–loving older brother, who prefers writing a passive-aggressive Yelp review to actual confrontation. While it can never quite be considered its own startup ecosystem, only in San Francisco can college tech interns party over the summer without parents having a clue. And only in San Francisco will people give you a look for walking into the office with your Starbucks at 11 a.m. (with Philz, Blue Bottle, and Sightglass across the street, you chose Starbucks?). Yet the talent in San Francisco is ridiculous: with the unicorns, ubercorns, and big business aplenty. San Francisco's local, organic charm draws the best of the best from across the world. And while it may be chided for its youth and immaturity, San Francisco is where youngsters thrive, thrown into a city where rules are seldom and startups are many; a city where entrepreneurs are not only born, but made.

VENTURE ACTIVITY Q4'15 - Q3'16

VC FUNDING	$18.2B
DEALS	1179
EXITS	220
5 YEAR YoY FUNDING GROWTH	24.6%
5 YEAR YoY DEAL GROWTH	7.6%

Source: *CB Insights*

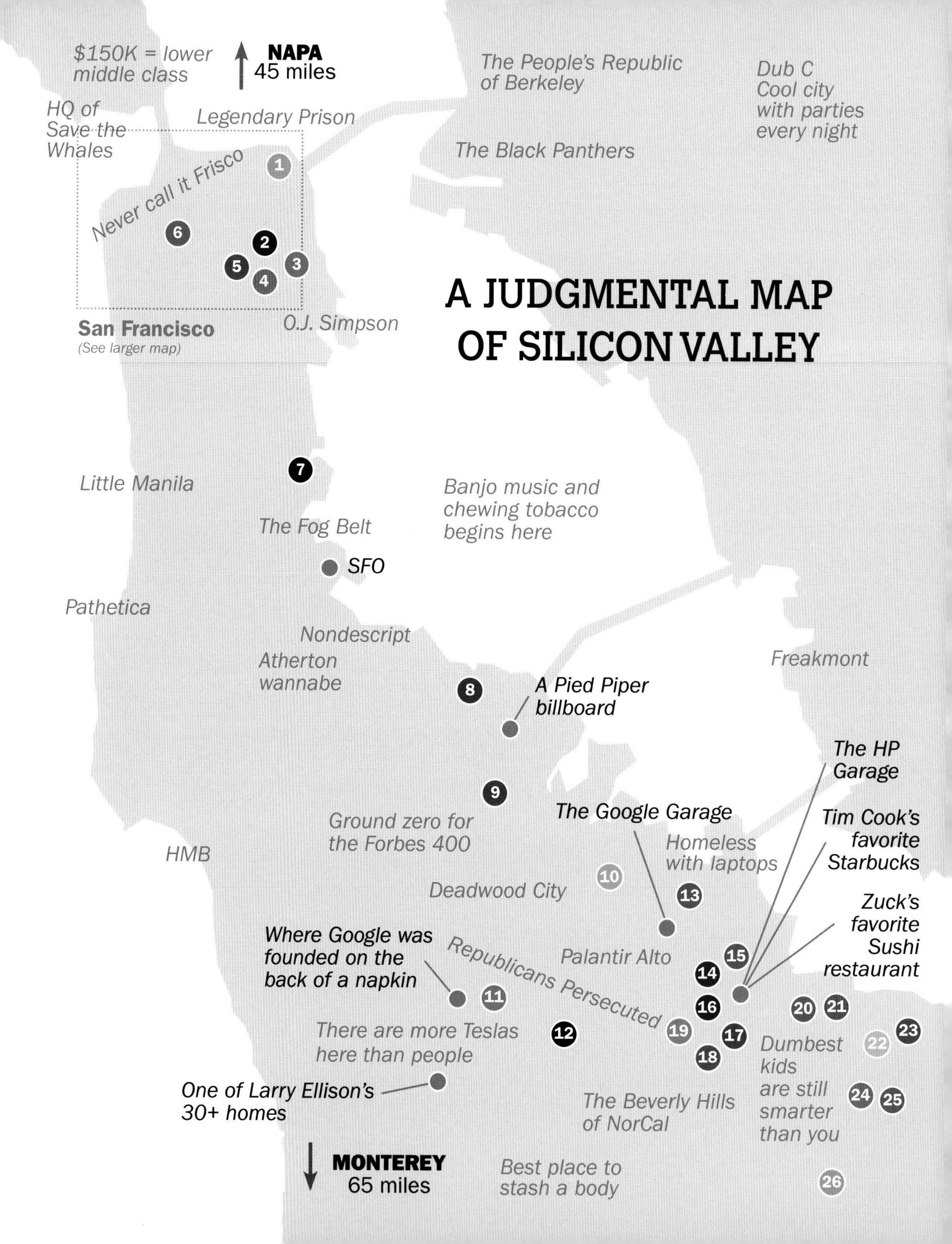
A JUDGMENTAL MAP OF SILICON VALLEY
$150K = lower middle class
NAPA 45 miles
The People's Republic of Berkeley
Dub C Cool city with parties every night
HQ of Save the Whales
Legendary Prison
The Black Panthers
Never call it Frisco
1
6
2
3
5
4
San Francisco
(See larger map)
O.J. Simpson
7
Little Manila
Banjo music and chewing tobacco begins here
The Fog Belt
SFO
Pathetica
Nondescript
Atherton wannabe
Freakmont
8
A Pied Piper billboard
The HP Garage
9
The Google Garage
Tim Cook's favorite Starbucks
Ground zero for the Forbes 400
Homeless with laptops
HMB
10
Deadwood City
13
Zuck's favorite Sushi restaurant
Where Google was founded on the back of a napkin
Republicans Persecuted
Palantir Alto
15
14
11
16
20
21
12
19
17
23
There are more Teslas here than people
Dumbest kids are still smarter than you
22
18
One of Larry Ellison's 30+ homes
24
25
The Beverly Hills of NorCal
MONTEREY 65 miles
Best place to stash a body
26

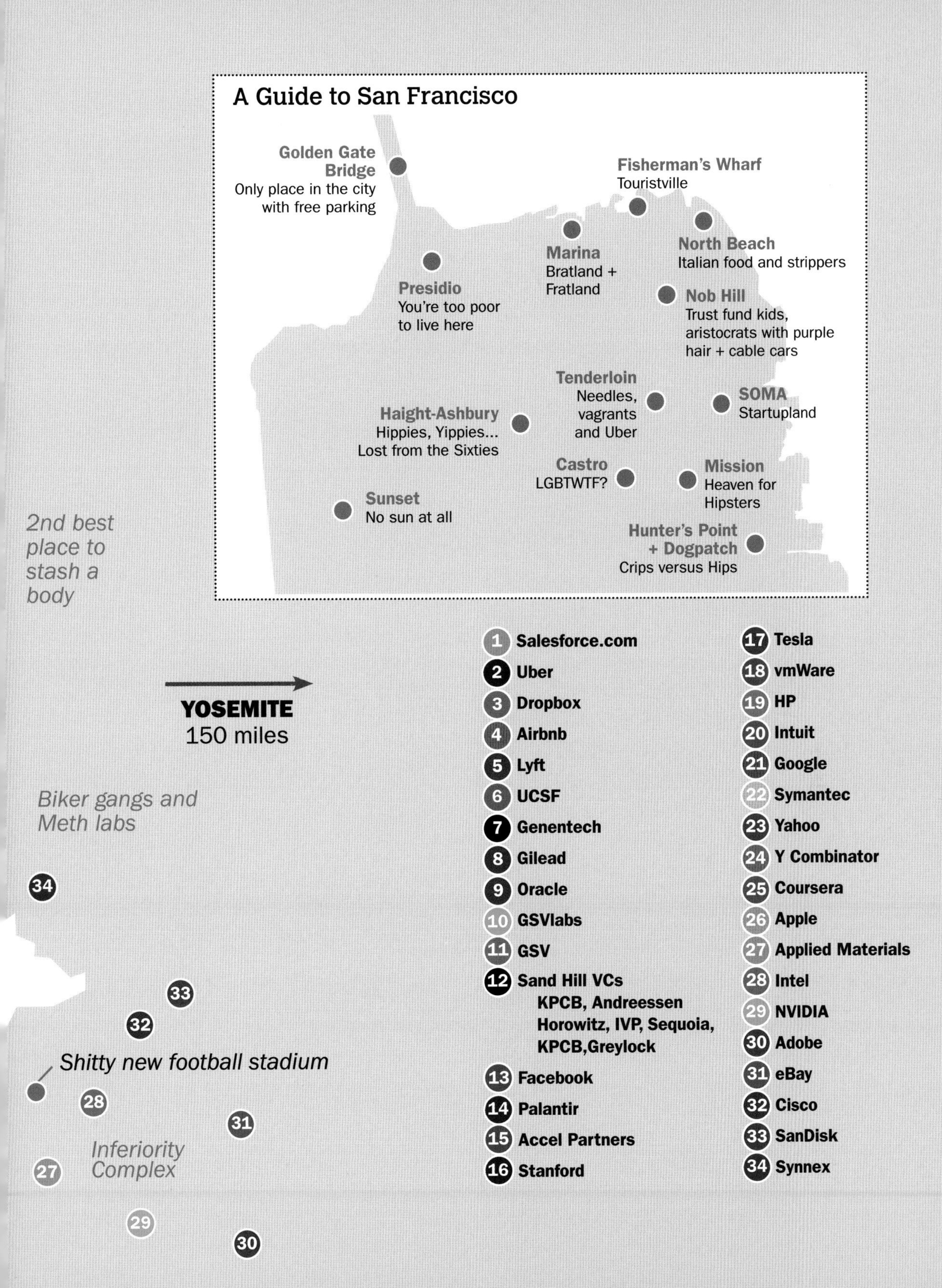
A Guide to San Francisco
Golden Gate Bridge
Only place in the city with free parking
Fisherman's Wharf
Touristville
Marina
Bratland + Fratland
North Beach
Italian food and strippers
Presidio
You're too poor to live here
Nob Hill
Trust fund kids, aristocrats with purple hair + cable cars
Tenderloin
Needles, vagrants and Uber
SOMA
Startupland
Haight-Ashbury
Hippies, Yippies... Lost from the Sixties
Castro
LGBTWTF?
Mission
Heaven for Hipsters
Sunset
No sun at all
Hunter's Point + Dogpatch
Crips versus Hips
2nd best place to stash a body
YOSEMITE
150 miles
Biker gangs and Meth labs
Shitty new football stadium
Inferiority Complex
34
33
32
28
31
27
29
30
1 Salesforce.com
2 Uber
3 Dropbox
4 Airbnb
5 Lyft
6 UCSF
7 Genentech
8 Gilead
9 Oracle
10 GSVlabs
11 GSV
12 Sand Hill VCs
KPCB, Andreessen Horowitz, IVP, Sequoia, KPCB,Greylock
13 Facebook
14 Palantir
15 Accel Partners
16 Stanford
17 Tesla
18 vmWare
19 HP
20 Intuit
21 Google
22 Symantec
23 Yahoo
24 Y Combinator
25 Coursera
26 Apple
27 Applied Materials
28 Intel
29 NVIDIA
30 Adobe
31 eBay
32 Cisco
33 SanDisk
34 Synnex

MO' MUNI MO' PROBLEMS

Some of the many not quite optimal transportation systems in San Francisco.

SF Municipal Railway
Muni for short—pronounced mew-knee—the public transit system of the city consisting of mainly trains, buses, and mediocrity.

Muni Bus
Make sure to have a couple dollars change in your pocket as you get ready to board the 44 local bus lines. You'll be able to traverse the city of SF, stopping at each and every crack in the sidewalk.

Muni Metro
A great option for the working entrepreneur on a budget, but its arrival time is as unpredictable as what kind of meat you'll find in a McDonald's McChicken sandwich.

Cable Cars
More on the expensive, touristy side, but a nice reminder of life before technology, skyscrapers, and optimization.

Uber/Lyft/Any other ride company with VC money to spend
Usually cheaper than the taxis.

Taxis
Not cheaper than the taxis.

ONE MAN'S **APARTMENT** IS A CASTLE IN ITALY

Average Apartment Rental **$3803/Month**
Median Apartment Sale Price **$1,130,000**
Median 1 Bed Apartment Sale Price . **$797,000**

Meanwhile...

Castle in Marche, Italy with watchtower and pool
$897,000

Fairy Island, Wisconsin (1-acre island)
$560,000

HOW TO BE A STEALTH **BANDWAGON GIANTS FAN**

1. **KNOW THE ROSTER.** And don't only know the roster, periodically quiz every supposed "Giants fan" about who that jersey number is, what that nickname is, etc. Madison Bumgarner? His name is "MadBurn" you poser, you just don't know the spirit of the game

2. **GET DRUNK AND YELL.** It could be about the game, about the ump, about your seats, about anything really. Nothing says passion like one too many beers and leaving in between every strike-out to relieve yourself.

3. **FABRICATE A STORY.** You need a reason to become a Giants fan. Maybe your dog died watching a Giants game. Maybe your father's college roommate's sister's husband's great-aunt was from San Francisco. Whatever the case, never let the die-hard fans who now have to pay exorbitant ticket prices know the actual reason: You needed something to believe in.

GSVlabs

In 1970, Xerox Corporation launched Xerox PARC in Palo Alto to create "the office of the future." Assembling a world-class team of experts in information technology and physical sciences, Xerox PARC was a catalyst for innovation—it yielded the original design for the personal computer (and famously inspired Steve Jobs to create the first Macintosh), the first graphical user interface, and even the computer mouse. And with that, Silicon Valley 1.0 was created.

Historically, Silicon Valley titans Apple, HP, and Google were all started in a garage, with the founders scrappily building their eventual multibillion-dollar companies from recycled metals and borrowed (or stolen) items inside of a dimly lit facility designed to store parked cars.

Fast-forward to today, and now, one-third of all startups that raised a Series A went through an accelerator. Why? The best explanation is if you want a job, create it yourself. Thus came the rise of startup incubators and accelerators. These communities provide would-be entrepreneurs with information, education, insights, resources, relationships, and—in some cases—capital to help bring their ideas to life. Budding college graduate–entrepreneurs don't have to work out of a garage to build their businesses but are now able to be with other like-minded crazies that are trying to do big things to change the world.

Founded in 2012, GSVlabs is a global innovation platform based in Silicon Valley that accelerates startups and connects corporations to exponential technologies, business models, and entrepreneurs. We believe this is a next-generation Xerox PARC. Welcome to Silicon Valley 3.0.

Connecting The World to Silicon Valley… And Silicon Valley to The World

At the core of GSVlabs is a community of game-changing entrepreneurs focused on key verticals, including big data, sustainability, education technology (EdTech), entertainment, and mobile. GSVlabs is home to over 170 startups that raised $200 million in 2015.

GSVlabs generates outsize economic value by accelerating both startups and corporate innovation, enhanced by a proprietary talent and advisory network. By combining these elements in a single platform, GSVlabs unlocks powerful network effects. Top entrepreneurs are typically drawn to an ecosystem with high-value resources and relationships, which in turn draws in global businesses and talent.

Silicon Valley remains the center of global innovation, with 15,000+ active startups and $27+ billion of venture capital deployed in 2015. But what's exciting is an emerging Global Silicon Valley, with $129 billion of VC investment worldwide in 2015.

The flagship event of GSVlabs, the Pioneer Summit, brings some of the biggest names in technology as speakers for the two-day event. Speakers at the summit included Bill Campbell, Anne Wojcicki, John Donahoe, Joe Lonsdale, Ann Miura Ko, Carol Bartz, Mike Abbott, Ron Johnson, Lila Ibrahim, Carlos Watson, Diane Greene, and Nikesh Arora, and startups featured at the summit have raised over $5 billion.

Michael Moe and Joe Lonsdale

THE GLOBAL SILICON VALLEY HALL OF FAME

Silicon Valley is not just about bits, bytes and chips. It's a mindset of innovation, growth and entrepreneurship. We created the Global Silicon Valley Hall of Fame to celebrate the Pioneers who created the foundation of Silicon Valley, which have changed the world for good.

These Pioneers inducted into the Global Silicon Valley Hall of Fame built a foundation in the 60 miles between San Francisco and San Jose that has transformed the world. Today, we're excited to be witnessing the dawn of a new era of global entrepreneurship and innovation. The values listed below have defined Silicon Valley and are being exported around the world to the leaders who are building tomorrow.

GSV HALL OF FAME INDUCTEES

Bill Campbell, Coach of the Valley
Carol Bartz, Fearless Builder
Diane Greene, Technology Maverick
Dick Kramlich, Champion of the Future
Gordy Davidson, Confidant of Champions
Ken Coleman, Pioneering Mentor
Larry Sonsini, Counsel to Titans
Andy Grove, Legendary Leader *(posthumously)*
Mike Homer, Creative Force of Nature *(posthumously)*

PAY IT FORWARD: People help you with the expectation that, in turn, you will help others, with no obligation or financial upside.
DREAM BIG: Whatever the mind can conceive and believe, it can achieve.
FAILURE = SUCCESS: Failing because you are lazy is never OK. But when you give it your all and it doesn't work, it's a growing process.
MERITOCRACY: It doesn't matter who your parents were, or where you went to school. What matters is how hard you work, how open you are the learn and how relentless you can be.
DON'T STOP THINKING ABOUT TOMORROW: Today's "disruptor" is tomorrow's "disrupted."
THINK DIFFERENTLY: If you don't want to change anything, come up with an idea that people agree with.

IN MEMORIAM **BILL CAMPBELL (1940-2016)**

It's impossible to overstate Bill Campbell's impact on Silicon Valley. He headed west to Palo Alto in 1982 to be a quota-carrying salesman at Apple Computer after being the head football coach at his alma mater, Columbia University. Nine months later, he was running sales, and in no time, Bill became the CEO of Claris Software, a key Apple division. In 1994, he was named the CEO of Intuit. But Bill's legacy wasn't forged until Kleiner Perkins' John Doerr inserted him as a "coach" to promising KPCB companies after noticing his incredible ability to cultivate leaders. Bill mentored Jeff Bezos at Amazon and Google's Eric Schmidt and Larry Page. A close friend and advisor to Steve Jobs, Bill served on Apple's Board for 17 years. What made Bill so special was his ability to make everybody feel like they were his best friend and that he cared deeply about them... because he was, and he did. We miss Bill every day but his spirit lives on and spurs us to change the world for good.

OFFICE **SPACE**

Silicon Valley offices are quite unlike others. Offices here are a cross between a fraternity house, spa, and playground. Oh, and don't be surprised to see graffiti artists spray-painting the walls, too.

1. Every playground and Silicon Valley startup needs a slide.
2. Engineers love donuts unless a donut becomes your startup's value.
3. No time for the great outdoors? We'll bring the great outdoors to you.
4. Art that hangs from the ceiling made with ping pong balls found under desks after endless beer pong tournaments.
5. A giant chess set, because ones that fit on tables are small time.
6. A gaming lounge and arcade, where you can blow off steam during breaks.
7. A plethora of alcohol in the form of full-stocked bars...How else do you think people get work done around here?
8. Ceiling with no ceiling tiles because it looks more "industrial."
9. Giant, beautiful wall murals that have the dual purpose of covering ugly paint jobs and serving as a great source of branding
10. Egg chairs—the best place where new ideas are hatched.
11. Executive's office or a six-year-old's dream playroom?
12. Dogs are often liked more and treated better than employees in the office.
13. Couches for those that need to powernap, i.e. recover from a hangover.
14. Mini putt-putt green, where employees can practice the sport of business.
15. Napping room where employees engage in intercompany hook ups.
16. Giant bean bag with an even larger teddy bear—a place where real men show their colors.
17. This West Wing replica is the closest most Silicon Valley executives want to get to the White House.

4

5
6

8
9

10

11
BarkBox
xtava

12
13

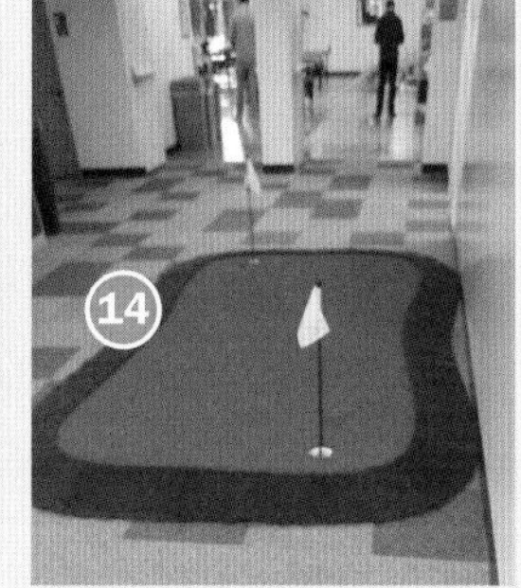
14

15

16

17

DIVERSITY IN SILICON VALLEY TECH GIANTS

	OVERALL				LEADERSHIP				TECH ROLLS			
	Women	Black	Hispanic	Asian	W	B	H	A*	W	B	H	A*
US	51%	13%	17%	5%								
Apple	30%	7%	11%	15%	28%	3%	6%	21%	20%	6%	7%	23%
Google	30%	2%	3%	30%	21%	2%	1%	23%	17%	1%	2%	34%
Twitter	30%	2%	3%	29%	21%	2%	0%	24%	10%	1%	3%	34%
Facebook	31%	2%	4%	34%	23%	2%	4%	19%	15%	1%	3%	41%
LinkedIn	39%	2%	4%	38%	25%	1%	4%	28%	17%	1%	3%	60%
eBay	42%	7%	5%	24%	28%	2%	2%	23%	24%	2%	1%	55%

Though called a meritocracy by some, Silicon Valley is still controlled by white males. While tech giants like Facebook and accelerators like GSVlabs are taking steps to correct this diversity gap, the problem persists. (Insert obligatory, politically correct, self-righteous, and outraged comment here.)

A* is Asians including Indians.

UNITED STATES **TOP ACCELERATOR LOCATIONS**

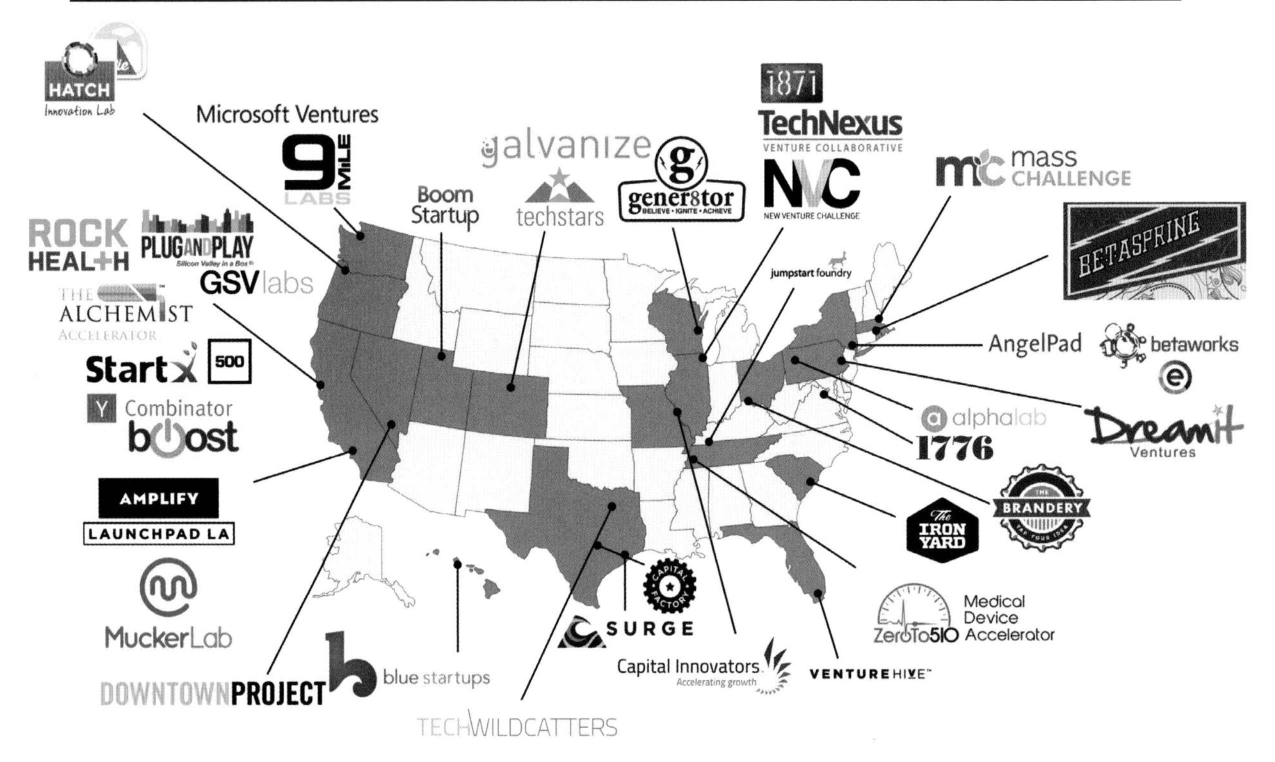

Required Reading for Entrepreneurs

Elon Musk, by Ashlee Vance
Finding the Next Starbucks, by Michael Moe*
Only the Paranoid Survive, by Andy Grove
Outliers, by Malcolm Gladwell
Stacking the Deck, by Dave Pottruck
Startup: A Silicon Valley Adventure, by Jerry Kaplan
Start-up Nation, by Dan Senor & Saul Singer
Steve Jobs, by Walter Isaacson
The Art of the Start, by Guy Kawasaki
The Career Playbook, by James Citrin
The Facebook Effect, by David Kirkpatrick
The Hard Things about Hard Things, by Ben Horowitz
The Innovator's Dilemma, by Clayton Christensen
The Lean Startup, by Eric Ries
The Monk and the Riddle, by Randy Komisar
The New New Thing, by Michael Lewis
The Nudist on the Late Shift, by Po Bronson
The Search, by John Battelle
The Silicon Boys, by David Kaplan
The Wisdom of Crowds, by James Surowiecki
The World Is Flat, by Thomas Friedman

*You can't love others if you don't love yourself.

QUESTION, WHICH ARTIST'S WORK IS WORTH THE MOST? PICASSO, RENOIR, MONET...

In 2005, Sean Parker, portrayed excellently by Justin Timberlake in *The Social Network*, president of Facebook at the time, came to mural artist David Choe with a proposition—paint his murals on the Facebook office walls for $60,000 or company stock. David painted the murals, which are still up today. When he was done, David chose to take $60,000 in stock instead of cash.

That was a smart decision because when Facebook IPO'd in 2012, David's Facebook stock was worth over $200 million. Assuming that he sold no shares of Facebook since then, David's stock is now worth over $600 million.

The Entrepreneur's
PLAYLIST

Song	Artist	Length
Start Me Up	The Rolling Stones	3:33
Living on a Prayer	Bon Jovi	4:10
Won't Back Down	Tom Petty	2:58
Dancing in the Dark	Bruce Springsteen	4:01
Stronger	Kelly Clarkson	3:41
A Milli	Lil Wayne	3:43
Story of My Life	One Direction	4:04
Eye of the Tiger	Survivor	4:05
We Will Rock You	Queen	2:02
C.R.E.A.M.	Wu-Tang Klan	4:12
Don't Stop Believing	Journey	4:09
Go Your Own Way	Fleetwood Mac	3:43
Firework	Katy Perry	3:47
I'm a Believer	The Monkees	2:46
Hall of Fame	The Script	3:22
All I Do Is Win	DJ Khaled	3:50
Revolution	The Beatles	3:25
People Like Us	Kelly Clarkson	4:19

FORGET FRIENDS WITH BENEFITS... #FRIENDSWITHZENEFITS

After Zenefits CEO Parker Conrad publicly resigned from his role amid turmoil at his high-flying startup, which seemed more like a fraternity than a company, the hangover started to kick in.

David Sacks replaced Conrad as CEO, who promptly acted to clean up the mess left behind. Sacks banned alcohol from the office, as well as access to the company's stairwells. The former makes sense, but why the latter? Well, it was reported that cigarettes, red Solo cups filled with beer, and used condoms were found in the stairwell.

What happens in Zenefits stays in Zenefits.

BENEFIT	f	VS. A TYPICAL STARTUP
FOOD	Gourmet food courts, private chefs, candy shop, micro-kitchens, all with gluten-free and allergy-free alternatives	Coke and Sprite in snack room if you get there before nine at the start of the month after Costco run
PARENTAL LEAVE	Four months of paid maternity/paternity leave	Bring baby to work, part-time intern babysits
TRANSPORTATION	Shuttle from San Francisco to headquarters, free bikes across campus, bike repair shop, electric car station	Alternate between mother and father dropping you off
BRANDING	Facebook store with sweatshirts, pants, stuffed animals, and far more	Hand out free Frisbees at college career fair
ENTERTAINMENT	Video game arcade, walls you can write on, expansive and high-quality gym, rock-climbing wall, silk-screen facility to vent out stress through art, music rooms	$25 3-in-1 Ping-Pong, foosball, air hockey table bought at garage sale; dreaming about what you'll do with all that equity
TECH HELP	IT help desks, computer accessory vending machines for forgotten chargers and adapters	Yahoo Answers, retired father-in-law at neighborhood BBQ who used to work at Cisco
SERVICES	Barber shop, guided tours	Water fountain, hand-made card on birthday

Here are the top names to follow on Twitter to stay plugged in to what's buzzing around the digital Silicon Valley*

Marc Andreessen @pmarca
If you're worried about robots eating all the jobs, maybe we should stop programming students as if they're robots

Cofounder, Andreessen Horowitz *600K+ Followers*
Tweeted over 100+ times daily. As of fall 2016, Marc is currently on a Twitter break, and we all eagerly wait for our favorite Tweeter to emerge.

Kara Swisher @karaswisher
Yer killing me Marissa. To be fair, that might be your goal.

Executive Editor, Recode *1M+ Followers*
Journalist who consistently breaks the top headlines in Silicon Valley. Yahoo's Public Enemy No. 1.

Elon Musk @ElonMusk
The rumor that I'm building a spaceship to get back to my home planet Mars is totally untrue.

CEO, SpaceX + Tesla *6M+ Followers*
Stirs up the Twittersphere with his occasional cheeky tweet. Despite what he says, we're 99% sure he's from Mars.

Danielle Morrill @DanielleMorrill
I look forward to the time when money is the blocker to my startup's problems. Right now it's still time.

CEO + Cofounder, Mattermark *50K+ Followers*
Startup founder that's Tweetstorm buddies with Marc Andreessen. When she's not ruling as the tech-savvy Queen of Twitter, she's baking.

Bored Elon Musk @BoredElonMusk
Mars now officially has more water than California.

Not The Real Elon Musk *650K+ Followers*
What Elon Musk really thinks. Full of ideas that are half crazy, half good.

Startup L. Jackson @StartupLJackson
If you're parents understand what your company does, you're probably not innovating.

Silicon Valley's Digital Yoda *80K+ Followers*
Pithy commentary on tech culture and investor behavior. He was Silicon Valley's own Gossip Girl, and everyone wondered who he was...until it was uncovered it was VC Parker Thompson.

THE TWEET HEARD AROUND THE WORLD

Silicon Valley professional tweeter Marc Andreessen made a quip about Facebook on Twitter, and offended 1 billion Indians with a single tweet.

He quickly apologized and then disappeared from his favorite social media platform altogether, leading techies to wonder—"Where did Marc Andreessen go?"

His whereabouts were unknown for two weeks, as not a peep was heard from him, unusual for the man who tweets usually 100+ a day.

All was well in the world again, when Andreessen emerged from his "social media vacation" to send out a congratulatory tweet. Whether or not he will continue his social media exile is currently unknown.*

*As of 15 minutes ago.

2

NEW YORK CITY

What once was the city that never sleeps is now Silicon Alley, a very established startup ecosystem that has less of a focus on "hard" technology and more of a focus on UI. This startup scene is rescuing people from the safer path of the lawyer/banker/consultant and instead putting them on the treacherous, risky, passion-fulfilling trajectory of the entrepreneur, though some employees would still rather have salary than equity (it's not quite the Valley yet). Benefiting from its proximity to fashion, financial services, and media as well as the hordes of angel investors looking for the next Tumblr or Etsy (rumor has it that renowned angel Joanne Wilson was ready to put a million dollars into Betsy, the Etsy for cows), New York has seen a huge spike in venture capital money that is likely to continue.

VENTURE ACTIVITY Q4'15 - Q3'16

VC FUNDING	$11.8B
DEALS	1258
EXITS	313
5 YEAR YoY FUNDING GROWTH	28.2%
5 YEAR YoY DEAL GROWTH	10.1%

Source: *CB Insights*

NEW YORK **AT A GLANCE**

Power Breakfast
- Balthazar
- Core Club
- Lafayette
- Loew's Regency
- The NoMad

Best Pizza
- Artichoke
- Joe's, Greenwich Village

Coffees Shops to Start Up In
- Ace Hotel
- Gregory's
- Joe's
- Ninth St. Espresso

Deal-Closing Dinner
- Locanda Verde
- Marea
- Per Se
- Peter Lugers

Bar
- Crosby Bar
- Dear Irving
- Jadis
- Le Bain
- Oldtown
- Raines Law Room

High-End Social Clubs
- CORE Club
- Soho House

Local Beer
- Brooklyn Brewery
- The Other Half

Go-to Shot
- Boilermaker
- Pickleback

Startup Districts
- Brooklyn Navy Yard
- Chelsea
- Dumbo
- Flatiron
- NoHo
- Union Square

Entrepreneurship Events
- CoFounders Lab
- Entrepreneur Week
- Lean Startup Machine
- NAPEC Innovation Conference
- New York Tech Meetup
- OZY Fusion Fest

"Must Be At" Local Events
- Central Park SummerStage
- Comedy Cellar's Late Night
- Manhattanhenge
- Movies with a View
- Shakespeare in the Park
- Tribeca Film Festival
- U.S. Open
- Yankees game

THE TOP

ANGELS
- Alexis Ohanian
- David Tisch
- Joanne Wilson
- John Katzman
- Jordy Levy

ACCELERATORS
- Betaworks
- Dreamit Ventures
- Entrepreneurs Roundtable Accelerator
- Fintech Innovation Lab
- First Growth Venture Network
- Techstars

VCs
- Bessemer
- General Atlantic
- Greycroft Partners
- Insight Ventures
- Lerer Hippeau Ventures
- RRE Ventures
- Union Square

Q4'15 - Q3'16 TOP FUNDING ROUNDS
- Oscar—$400M
- Jet—$350M
- CommonBond—$300M
- Athena Art Financial—$280M
- Karoo—$250M
- OTG—$250M

GIRL POWER

The New York startup ecosystem benefits from a host of women entrepreneurs and investors helping to close the gender gap in tech.

KATHRYN MINSHEW *CEO, Muse*

Kathryn is the Muse's CEO and number one swashbuckler. Kathryn has spoken at MIT and Harvard, appeared on The TODAY Show and CNN, and contributes on career and entrepreneurship to the Wall Street Journal and Harvard Business Review. She was named to *Forbes*' 30 Under 30 in Media and Inc.'s 15 Women to Watch in Tech. Before founding The Muse, Kathryn worked on vaccines in Rwanda and Malawi with the Clinton Health Access Initiative and was previously at McKinsey.

JOANNE WILSON *Angel Investor, Co-Founder, Women's Entrepreneur Festival, Gotham Gal Ventures*

Joanne got her start in the tech scene working on the media side of the technology industry during the 1990s dot-com boom. She made her first angel investment in 2007 and since then has accumulated a portfolio of over 90 companies, focusing her thesis on female founders. She created The Women's Entrepreneurs Festival (WEFestival), which brings together women entrepreneurs to connect and be inspired. She is an insatiable foodie, reader, traveler, art collector and music connoisseur.

RESHMA SAUJANI *Founder and CEO, Girls Who Code*

A lawyer and politician by day, Saujani is the founder of Girls Who Code, dedicated to "inspire, educate, and equip girls with the computing skills to pursue 21st century opportunities." Girls Who Code provides intensive summer programs for high school girls, with Twitter, Facebook, Google, and many more companies as partners. Saujani wants to teach a million women how to code by 2020.

SHANA FISHER *Managing Partner, High Line Venture Partners*

Though venture capital firms often feel like an old boys clubs, Fisher founded her own firm, High Line Venture Partners, and serves as its managing partner. The firm has invested in companies like Aereo, imgix, and PillPack. As an angel investor, Fisher's crowning achievement was a $500,000 angel investment in a little-known company at the time: Pinterest. Pinterest is now valued at $11 billion.

TOP CITIES FOR FEMALE ENTREPRENEURS BY WOMEN AS A PERCENTAGE OF STARTUP FOUNDERS

City	%
Chicago	30%
Boston	29%
Silicon Valley	24%
SoCal	22%
Montreal	22%
Paris	21%
Tel Aviv	20%
Toronto	19%
Singapore	19%
Kuala Lumpur	19%
London	18%
Moscow	17%
New York	17%

COMMONLY SEEN **IN NEW YORK...**

All the stereotypes of New York City in one uplifting page.

The Taxi Driver with Anger Management Issues ←
Cursing his friends, cursing his family, cursing his passengers, but mostly cursing Uber and Lyft.

The Artsy NYU Student →
The skateboarder/fashionista/photographer/filmmaker with more Instagram followers than dollars in her savings account.

Cigarette-Smoking Hot Dog Guy ←
Advertising two products that will bring you a slow, painful death.

The Tourist ↘
See "obstacle" or "slow, stupid blob that doesn't get the hell out of my way."

Street Musicians ↘
More talented than the makers of the Starbucks albums you buy, but do not offer the Pumpkin Spice Latte in the Fall...sorry ladies.

The Drugged-Out Wall Street Wolf ←
Though he looks more like your balding uncle than Leonardo DiCaprio, he will slam you over the head with his briefcase full of cocaine if he doesn't get his hot dog in under 30 seconds.

COWORKING **SPACES**

With absurdly little space and a high concentration of wealth, the price of real estate in New York is just as unreal as it is in San Francisco. Enter coworking spaces, with multiple startups sharing a single space for cheap, paying by the desk, and earning perks like gourmet coffee, Yoga Tuesdays, and 3-D printers. Here are three of the hottest coworking spaces in the city.

WeWork Bryant Park

The company with a $16 billion valuation has locations everywhere from Chelsea to SoHo. On track to become the fastest-growing leaser of new office space in America, WeWork is building a community where similarly minded people can do everything together—work, eat, sleep, exercise, build a business...talk about a sharing economy.

General Membership. . . $45/month
Hot Desk $220/month
Dedicated Desk $325/month
Private Office $450/month

Hot desks are guaranteed but may be different each day. Dedicated desks are assigned and personal.

Fueled Collective SoHo

On the way to Fueled Collective, you'll pass notable startups like Thrillist and FourSquare as well as the many Prada and Versace "doormen" who walk and talk like bouncers and never give you directions. This collective offers perks like a year-round ice cream bar, a Ping-Pong table, and a celebrated "dope chillout couch."

Hot Desk $450/month
Dedicated Desk $800/month

Projective Space Lower East Side

Open to all sorts of founders and professionals, Projective Space is a more curated community that was once home to Uber and Eventbrite. It also provides more professional perks like HR help and cloud infrastructure. In addition, they host weekly events with NYC startup CEOs as well as workshops on personal wellness and late-night karaoke jam sessions.

General Membership. . . $95/month
Hot Desk $225/month
Dedicated Desk $555/month

5-MINUTE VS. **5-STAR**

When you need fast food on your five-minute lunch break.

Delis You'll find them on every corner, but look out for gems like Katz's. I'll have what she's having.

Seamless A popular way to deliver food for businesses, Seamless is a part of the GrubHub collective and noted for being very easy to expense on the company.

Vendors Fox reported tourists buying $30 hot dogs. Don't buy a $30 hot dog.

When you need fancy food in your Ferrari.

Le Bernardin Midtown, French seafood
Michelin Rating ★★★
Specialty Dishes: black bass, lobster lasagna, tuna, and foie gras

Eleven Madison Park Flatiron, American
Michelin Rating ★★★
Specialty Dishes: seasonal tasting menu

Chef's Table at Brooklyn Fare Brooklyn, American,
Michelin Rating ★★★
Specialty Dishes: foie gras, caviar, seafood, raw fish

3

SILICON BEACH and LOS ANGELES

There's no better way to escape the Silicon Valley bubble than to come to the warm sands of Silicon Beach, where kids are founding startups like they used to found rock bands. Silicon Beach is both incredibly huge and incredibly diverse—from the many elite universities to Beverly Hills to Compton, there is every kind of person here, and this diversity is the area's greatest strength. For example, the Silicon Beach benefits from its proximity to Hollywood, having some of the leading digital video startups as well as celebrity endorsements for companies across the board. With less engineering talent but more of an understanding of mass-market practices, Silicon Beach founders (and their surfboards in the office: a different kind of board meeting, if you will), speak the language of art and advertising instead of C++ and Java. Even if it might be faster to go from San Francisco to LA than from LA to LA, companies like Snapchat, Tinder, Oculus VR, and Beats call Silicon Beach home for a reason.

AGGREGATE VENTURE ACTIVITY Q4'15 - Q3'16

VC FUNDING	$13.2B
DEALS	939
EXITS	309
5 YEAR YoY FUNDING GROWTH	32.3%
5 YEAR YoY DEAL GROWTH	8.2%

Source: *CB Insights*

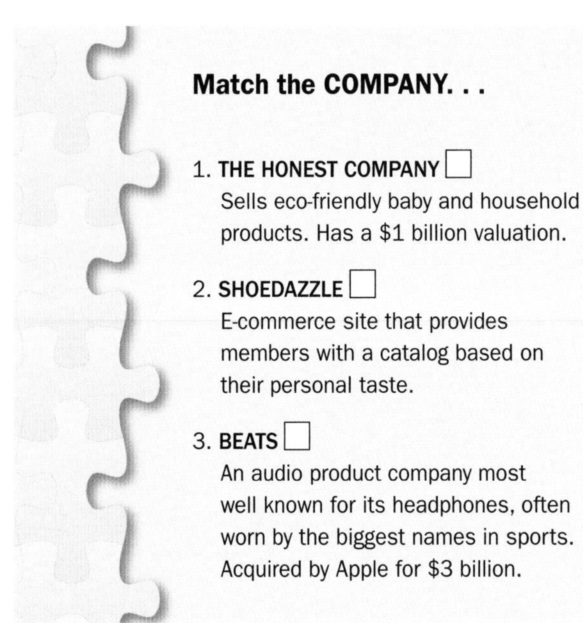

Match the COMPANY. . .

1. **THE HONEST COMPANY** ☐
   Sells eco-friendly baby and household products. Has a $1 billion valuation.

2. **SHOEDAZZLE** ☐
   E-commerce site that provides members with a catalog based on their personal taste.

3. **BEATS** ☐
   An audio product company most well known for its headphones, often worn by the biggest names in sports. Acquired by Apple for $3 billion.

. . . with the CELEBRITY FOUNDER

A. KIM KARDASHIAN
Really not sure what she's famous for, but she does have 33.7 million Twitter followers.

B. DR. DRE
Rapper who happened to earn his MD in West Coast G-Funk.

C. JESSICA ALBA
Actress known for Academy Award–winning roles in *Love Guru* and *Fantastic Four: Rise of the Silver Surfer*. Just kidding.

1: C; 2: A; 3: B.

SNAP: A brief, ephemeral, ten-second story.

SPRING 2011
When cofounder Reggie Brown says, "I wish these photos I am sending this girl would disappear," Evan Spiegel thinks it's a brilliant idea. The two are Kappa Sigma fraternity brothers at Stanford, and Spiegel brings in recent grad and fellow Kappa Sig Bobby Murphy to work on their project.

FALL 2011
Having spent the summer working on their app "Picaboo," with Spiegel as CEO, Murphy as CTO, and Brown as CMO, Brown is pushed out of the company after demanding 30% equity, something Spiegel and Murphy disagree with. By the start of fall, it's a two-man show titled "Snapchat."

APRIL 2012
With widespread popularity among high school students, the company reaches 100,000 users. Lightspeed Ventures' Jeremy Liew puts in $485,000 in a seed round for the company after hearing about the app from fellow partner Barry Eggers's high school daughter. Spiegel drops out of Stanford.

DECEMBER 2012
Mark Zuckerberg flies over to Southern California to meet with Spiegel and Murphy. There, he announces "Poke," Facebook's version of Snapchat. The founders return to their office and buy Sun Tzu's *The Art of War* for each of their six employees.

A TINDERELLA **STORY**

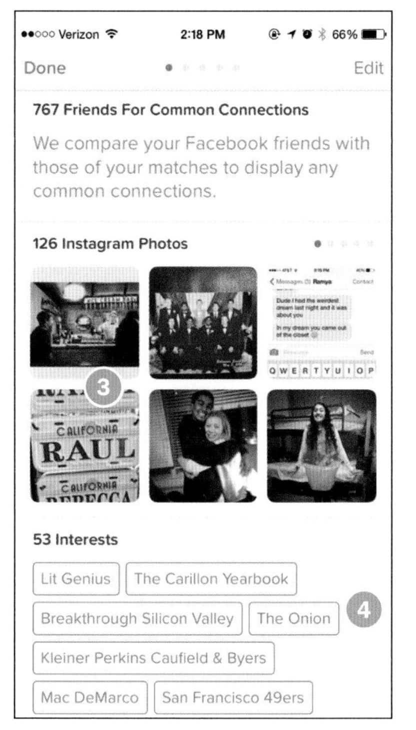

In the hectic life of an entrepreneur, LA-based company Tinder offers the chance to meet someone—a Tinderella story of sorts. Some tips to navigating the Tinder scene.

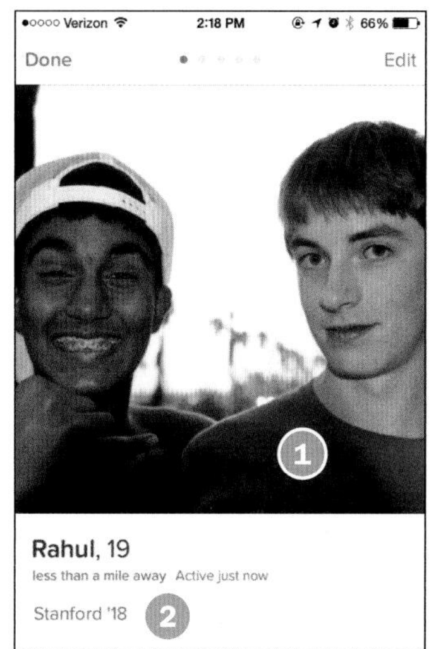

1. PICTURE Every picture of you should be of you and your very attractive, sexy friend. Gotta keep 'em guessing.

2. BIO There are two options for the bio. The first is the simple college name and year—remember, however, that it is unlikely you are sixty-five and part of the Stanford Class of 2018. The second is the generic—you like wine perhaps, or running, or pandas, or pandas running with wine.

3. INSTAGRAM Linking to your Instagram immediately gives you a boost of credibility. Having three filters applied to every single one of your photos dips that credibility, and having over 2000 photos destroys it.

4. ATTRACT Have a member of the gender you are attempting to attract fill this out for you, promptly Google all of these subjects as the matches start coming, and soon you'll be the Steph Curry at the three-point line.

NOVEMBER 2013
Snapchat launches their "Story" functionality, which proves popular among users. Zuckerberg offers Spiegel $3 billion for the app, but is promptly refused.

MAY 2014
Controversial e-mails are leaked from Spiegel's fraternity days, which paint a misogynistic picture of him.

FEBRUARY 2016
Gatorade's Super Bowl Snapchat filter gets 160 million impressions... more coverage than the game itself got.

NOVEMBER 2016
Snap launches Spectacles in disappearing vending machines and files for it's IPO.

LOS ANGELES AT A GLANCE

Power Breakfast
- Farm Shop
- Fig and Olive
- Gjelina
- Huckleberry
- Republique
- SQIRL

Coffee Shops to Start Up In
- Alfred Coffee
- Blue Bottle
- Coffee Commissary
- Espresso Cielo
- Menotti's
- Philz

Go-to Drink
- Boyd and Blair Potato Vodka

Offline Tinders
- 41 Ocean
- Bungalow
- The Room
- Zanzibar

"Must Be At" Local Events
- Coachella
- D23
- E3
- Film festivals
- LA County Fair
- UCB Asssscat Shows
- US Open of Surfing

High-End Social Clubs
- Jonathan Club
- SoHo House

Startup Neighborhoods
- Downtown LA
- Santa Monica
- South Bay (El Segundo area)
- Venice

LA Networking Tips
- Getting funded: Wear clean clothes with no tattoos on display.
- Drop more titles than names.

Top Celebrity Venture Capitalists

From Ashton Kutcher to MC Hammer, tons of celebrities have poured their dollars into investing in the hottest businesses that they believe in. In fact, Madonna loved Vita Coco so much that she invested in them. Here's a list of the most notable celebs in VC.

Ashton Kutcher, Actor
Invested in: Airbnb, Chegg, Skype, Casper, Fab

Bono, Musician, U2
Invested in: Facebook, Yelp, Forbes, Palm

Leonardo DiCaprio, Actor
Invested in: Mobli, Diamond Foundry, Rubicon Global

Jared Leto, Actor and Musician
Invested in: Blue Bottle Coffee, Nest, Meerkat, reddit

Nas, Rapper
Invested in: Coinbase, Genius, DeviantArt

Justin Bieber, Musician
Invested in: Spotify, Tinychat, Stamped, Shots

Magic Johnson, Former Basketball Player
Invested ownership in: Starbucks, Los Angeles Dodgers, Fast Food chains, Sodexo

MC Hammer, Rapper
Invested in: Square, Bump, Pandora, Flipboard

Kobe Bryant, Former Athlete
Recently launched $100 million venture capital fund.

Tyra Banks, Former Supermodel
Investments in: The Muse, Videogram, The Hunt

REJECTED!

In 2013, Snapchat rejected a $3 billion buyout offer from Facebook and a $4 billion offer from Google. The widely criticized decision paid off, with Snapchat now valued at $16 billion. Here are some of the other largest rejected buyout deals.

Facebook

Company Interested: **Yahoo!**
Offer: $1 billion (2006)
Market Cap: $350+ billion

Groupon

Company Interested: **Google**
Offer: $6 billion (2010)
Market Cap: $2+ billion

Yahoo!

Company Interested: **Microsoft**
Offer: $44.6 billion (2008)
Market Cap: $40+ billion

Twitter

Company Interested: **Facebook**
Offer: $500 million (2008)
Market Cap: $13+ billion

THE TOP

ANGELS

- Brad Jones
- Chris Sacca
- Guy Oseary
- Mark Suster
- Paige Craig

VENTURE CAPITALISTS

- CAA Ventures
- Clearstone
- CrossCut Ventures
- March Capital Partners
- Rustic Canyon Ventures
- Upfront Ventures

ACCELERATORS

- Amplify
- Disney
- IdeaLab
- LaunchPad LA
- Media Camp
- MuckerLab
- Science Inc.
- TYLTLab

ENTREPRENEURSHIP EVENTS

- CoFoundersLab Matchup Santa Monica
- Milken Institute Global Conference
- Santa Monica New Tech: Startup Pitchfest
- Startups Uncensored
- Techout LA
- The Montgomery Summit
- The Startup Conference

BIG SUCCESS STORIES

- BEATS
- Cornerstone OnDemand
- Rubicon Project
- Snapchat
- SpaceX
- The Honest Co.

Q4'15 - Q3'16 TOP FUNDING ROUNDS

- Snap—$1.8B
- Relativity Media—$400M
- Pharmaron Holding—$280M
- William Morris Endeavor—$250M
- OTG Management—$250M

A GUIDE TO **SNAPCHAT**

If you're still sharing pictures with your friends using e-mail, you're doing it wrong. Extremely popular with millennials, Snapchat is the platform to use to chat by snaps.

BREAKING DOWN THE SNAP

1 You can draw on, add emojis or text to your photos using these buttons.

Pro-tip: *Don't overdo this. Remember, this is Snapchat... not Crapchat.*

2 Snapchats are timed so your friends can view them only for a few seconds. After that, they "disappear."

3 Save your photo or video and share it on other platforms like Twitter, Instagram, and Facebook.

4 Add your Snap to your Snapchat Story and share it with the Snapchat community. Don't know what a Story is? More on that below.

5 Once you're satisfied with your photo, click here to send it off to your friends. If you don't have friends on Snapchat, make some. Social media requires being social.*

*Future employability warning: iPhone-savvy Snapchatters can screen shot and save any Snapchat you send.

SNAPCHAT STORIES

Snapchat Stories are a collection of Snapchat videos and photos taken that live on the platform for 24 hours.

Pro-tip: *Fake it until you make it. A great Snapchat Story is the fastest way to seem cool. Even if your life is boring, do something interesting for one minute.*

SNAPCHAT DISCOVER

Snapchat's Discover is becoming one of the best ways to consume the news. Hate reading boring articles? Snapchat Discover is for you. A few swipes right, down, and up and you'll know everything that's happening from your attic to the Arctic.

SAN DIEGO

While LA and San Diego are often grouped together, San Diego detests the comparison, believing itself to have better beer, better beaches, better burritos, and more important, better biotech. With UC San Diego's stellar biomedical engineering programs as well as the biotech industry with companies like Illumina getting hotter by the minute, San Diego is in the running as the biggest life science hub in the United States. This begs the question: Which will come first, the collapse of the biotech stock bubble or the creation of a super species of genetically enhanced superhuman X-Men? Given the hits and misses of biotech, it's no coincidence that San Diego is also the craft beer capital of America.

SAN DIEGO AND LA JOLLA VENTURE ACTIVITY Q4'15 - Q3'16

VC FUNDING	$1.9B
DEALS	189
EXITS	71
5 YEAR YoY FUNDING GROWTH	6.6%
5 YEAR YoY DEAL GROWTH	5.3%

Source: *CB Insights*

DIVISIONS OF BIOTECH

Like all sciences, biotechnology is a huge field and broadly defined, but can generally be thought of as applying technology to mess with biological processes. Some different kinds of biotech include:

Bioinformatics

Understanding and analyzing data from biology.
SD Company: GlySens
Total Funding: $12M

Biopharmaceutical

Medicines made from biological sources.
SD Company: aTyr Pharma
Total funding: $180.5M

Agricultural

Oldest biotech which involves altering organisms for some effect, often maximizing production.
SD Company: Cibus Global
Total funding: $42.5M

PUBLIC FEARS CONCERNING BIOTECH

Human Mutations

First, you're eating GMO corn, then your DNA gets sequenced, then your baby has a government-mandated third arm to increase efficiency, then you become GMO corn.

Bubble Collapse

As biotech stocks reach new heights, some warn of an impeding bubble while others say growth is justified. We say...if it's bioengineered, won't the bubble be longer lasting, able to grow in all environments, and bulletproof?

Radioactive Animals

The only thing worse than a squirrel that isn't scared of you and tries to take your food is a squirrel with three eyes that isn't scared of you and tries to take your food.

SAN DIEGO AT A GLANCE

Power Breakfast
- Claire's
- L'Auberge
- Westgate Hotel

Coffee Shops to Start Up In
- Claire De Lune
- Filter
- Lestat's

Deal-Closing Dinner
- Cucina Urbana
- George's at the Cove
- Market
- Pamplemousse

Entrepreneurship Events
- Demo Night
- SD Entrepreneur Day
- Startup Weekend SD
- UCSD Entrepreneur Challenge

Watering Holes
- Ballast Point
- Building K
- Mission Brewery
- Societe

ASU GSV SUMMIT

The annual ASU GSV Summit is the preeminent education innovation summit in the world. The past year's summit included 4,000+ attendees and 350+ EdTech Companies, with keynotes from people like Richard Branson, Howard Schultz, Mike Milken, Common, Reed Hastings, Magic Johnson, and Bill Gates.

THE TOP

ANGELS
- Allison Long Pettine
- Doug Hecht
- Howard Lindzon
- Taner Halicioglu
- Tim Reuth

VCs
- ARCH Venture Partners
- Avalon Ventures
- Correlation Ventures
- Domain Associates
- Qualcomm Ventures

ACCELERATORS
- CyberHive
- EvoNexus
- HERA LABS
- JLABS
- Moxie Foundation
- startR

Q4'15 - Q3'16 TOP FUNDING ROUNDS
- Human Longevity—$220M
- Sorrento Therapeutics—$150M
- PaxVax—$105M
- KnuEdge—$100M
- MobiMagic—$100M

SILICON VALLEY HISTORY

It is in the nature of entrepreneurs to always be moving, to think of the future, to build the next iteration, to market the next prototype. However, Silicon Valley historian Leslie Berlin, author of *The Man Behind the Microchip: Robert Noyce and the Invention of Silicon Valley*, shows the value of looking at the past to help understand the present in an exclusive interview.

How did Fairchild Semiconductor influence Silicon Valley?

Fairchild was very important. It was the first successful silicon and influential company in Silicon Valley. In fact, nearly all of the chip companies can trace their roots back to Fairchild. For example, Eugene Kleiner, who was one of the founders of Fairchild, cofounded the early venture capital company Kleiner Perkins Caufield and Byers. Similarly, Don Valentine, the founder of Sequoia, was a very high-level marketing guy at Fairchild. There are connections to Apple as well. Arthur Rock, who was a key backer of Apple, was the man who convinced Sherman Fairchild to start Fairchild. Mike Markkula, who basically cofounded Apple, introduced Jobs to Wozniack, was at Fairchild and then became an Intel guy.

It's not just who worked at Fairchild that was significant. On a cultural level, too, they were among the first to push for what is considered now the very foundations of Silicon Valley culture: stock options, flat organizational structures, no assigned parking spaces, uniform office spaces, and casual dress codes. For all of that, Fairchild was key.

Fairchild was astronomical—but like a compressed neutron star, the whole thing fell apart. **If Fairchild had managed to stay together, its influence might not (ironically) have been as great. Just think, if they had managed to stay together, then that talent would have all stayed concentrated.**

What's been the dark side of the Valley?

Back when chips were made here, there were all kinds of sweatshops, all over the Valley, where minority women did assembly work for almost no money. There was also, for a long time, a concern over high divorce rates, families falling apart, and the workaholic lifestyle. The environmental damage to the Valley, particularly from the era of

manufacturing, also really shocked people—what was so appealing to the notion of high-tech was that there were no smoke stacks. **No one was trying to do anything bad; people just didn't know. That was kind of a sad moment. It didn't come free.**

How has the crash of the 2000s shaped your current view of silicon valley?

That was an interesting time because there were people who were flying high, and then they were suddenly moving back to the Midwest to live with their parents—Harvard MBAs were out of work.

History is written by the winners—it's very difficult to find accurate records of failure because they tend to disappear. The dot-com era made these failures very real for the first time. For every company that made it through to the other side, hundreds just didn't. The experience also made us realize how sometimes it just comes down to luck. I knew two people who were working at the same company and they were equally smart and had similar amounts of company stock, but there was one person who was buying shares on margin when the bubble burst, and they ended up in terrible debt; meanwhile, the other person just happened to sell at the right moment, and now that person is a gazillionaire. **It was like playing a game of musical chairs. No one knew when the music was going to stop. How you fared just depended on whether you had already grabbed a chair.**

Do you think Silicon Valley culture has changed?

It has definitely changed from the beginning to where we are now. When I started studying early Silicon Valley, I would read about people saying things like, "We were never in it to make money," and I didn't believe it. But when I started reading letters these trailblazers would write home, I began to realize it was true!

Back then, a lot of the entrepreneurs were first and foremost academics and scientists who just got so excited by the idea of creating something new. But when someone said, "I'll give you some money to turn that idea into a company," then boom! There was a company made—a company that could create things a lot more than researchers. When they realized they were going to be rich, they were of course very excited and really happy, but it wasn't the reason they went into it.

Now, everyone goes into entrepreneurship with the understanding that getting rich is the desirable outcome. It might not be the only outcome, it might not be the only motivation, but **I think that there was a certain naiveté about the kind of fortunes that could be made back then that doesn't exist anymore.** Everyone is in on the secret.

What are some lessons entrepreneurs now can learn from the history of Silicon Valley?

One thing that was recognized in the early going was the importance of mentors, which isn't recognized as much now. Another important lesson is having the ability of having confidence in your own ideas while showing a willingness to listen to other ideas. If you don't have enough confidence, you can easily be swayed, and never accomplish anything in the end. If, on the other hand, you're completely locked down with your own idea, then the best-case scenario is that you alienate your employees and they leave; the worst-case scenario is that you alienate your employees, and they leave, and they were right, and your company goes down.

A third piece of advice originates from Robert Noyce, founder of Intel, who recognized that: **You can't do it all.** Noyce began as one of the eight founders of Fairchild, and by the end of it, he had all the reins in his hands. When he realized he couldn't run such a large company well, he cofounded Intel with Gordon Moore. He shared the executive suite, but ran a successful and groundbreaking company. Working with great people to do even greater things is so important here in the Valley, which I think is one of the biggest takeaways of the Silicon Valley culture.

Further HISTORICAL reading

The Man Behind the Microchip, Leslie Berlin
Regional Advantage, AnnaLee Saxenian
Cities of Knowledge, Margaret O'Mara
Secrets of Silicon Valley, Deborah Piscione

THE PHOTO HISTORY OF SILICON VALLEY

A picture says a thousand words. Here are six landmarks of the region.

1. **THE HP GARAGE** in Palo Alto, where Dave Packard and Bill Hewlett started their business, Hewlett Packard, in 1938.

2. **A SIGN COMMEMORATING THE HP GARAGE,** known as the Birthplace of Silicon Valley.

3. **THE APPLE GARAGE** in Los Altos, where Steve Jobs and Steve Wozniack built the first Apple computers.

4. **FACEBOOK'S HAND,** in front of their headquarters in Menlo Park. Also a place where you'd find the most Japanese tourists at any time in California.

5. **THE STANFORD DISH,** a radio telescope on the Stanford Foothills, which is a nice, breezy 3.5-mile hike altogether. Many of the top Silicon Valley players power walk this hike regularly.

6. **GOOGLE'S INFAMOUS BIKES,** found on their campus in Mountain View. They're free to take, so you can spot these bikes in Mountain View, Palo Alto, San Jose, Los Angeles, and probably even Mexico.

1
2
BIRTHPLACE OF "SILICON VALLEY"
THIS GARAGE IS THE BIRTHPLACE OF THE WORLD'S FIRST HIGH-TECHNOLOGY REGION, "SILICON VALLEY." THE IDEA FOR SUCH A REGION ORIGINATED WITH DR. FREDERICK TERMAN, A STANFORD UNIVERSITY PROFESSOR WHO ENCOURAGED HIS STUDENTS TO START UP THEIR OWN ELECTRONICS COMPANIES IN THE AREA INSTEAD OF JOINING ESTABLISHED FIRMS IN THE EAST. THE FIRST TWO STUDENTS TO FOLLOW HIS ADVICE WERE WILLIAM R. HEWLETT AND DAVID PACKARD, WHO IN 1938 BEGAN DEVELOPING THEIR FIRST PRODUCT, AN AUDIO OSCILLATOR, IN THIS GARAGE.
CALIFORNIA REGISTERED HISTORICAL LANDMARK NO. 976
PLAQUE PLACED BY THE STATE DEPARTMENT OF PARKS AND RECREATION IN COOPERATION WITH HEWLETT-PACKARD COMPANY, MAY 19, 1989.
THIS PROPERTY HAS BEEN LISTED IN THE
NATIONAL REGISTER
OF HISTORIC PLACES
BY THE UNITED STATES
DEPARTMENT OF THE INTERIOR
3
4
facebook
1 Hacker Way

5

6

A TIMELINE OF **SILICON VALLEY**

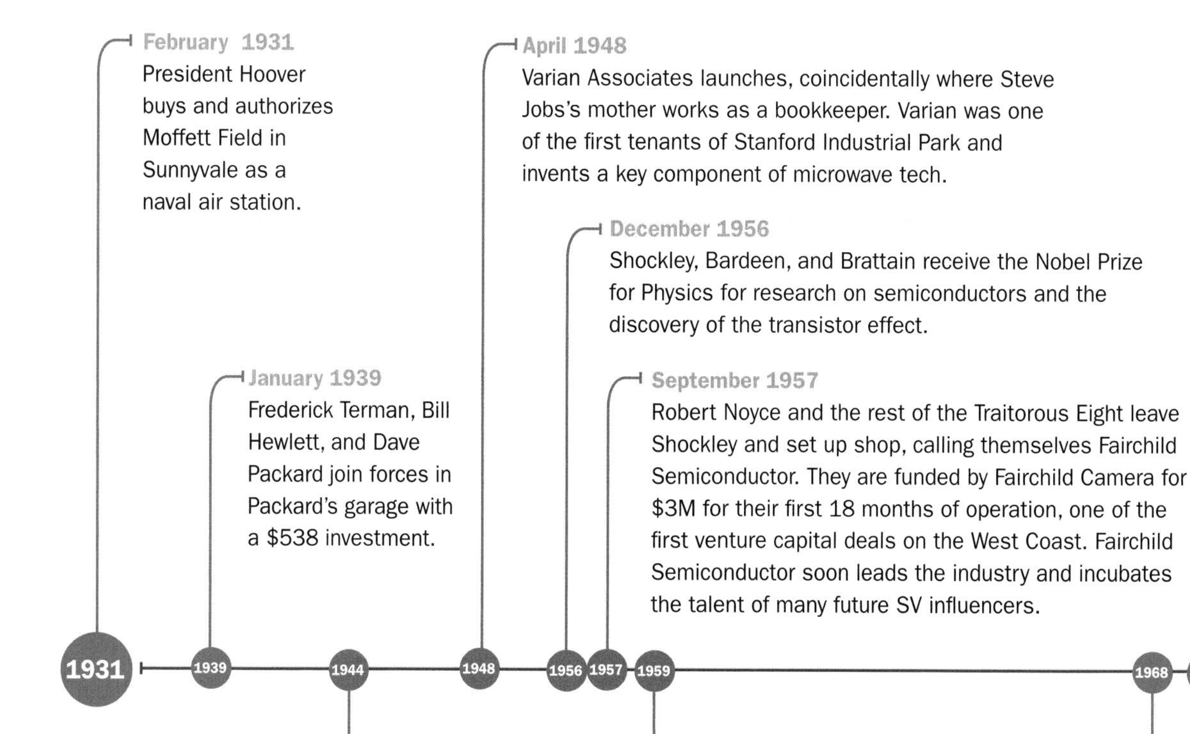

January 1944

Frederick Terman is appointed Stanford's Dean of Engineering (and later appointed provost from 1955 to 1965). Terman is angered that the U.S. government's $450 million (in 1945 dollars) fund for R&D gives Stanford only $50K, while Caltech receives $83M and MIT receives $117M. He immediately develops Stanford Engineering by recruiting talent from the East Coast, including hiring 11 members of the prestigious Harvard Radio Research Lab. He also encourages his best students to start their own companies rather than get PhDs, which transforms the Valley.

September 1959

Draper, Gaither, and Anderson becomes the first venture capital firm in Silicon Valley, setting a precedent for modern VC firms' partnership structure, division of profits, and focus on profits as an end in themselves.

JULY 1968

Robert Noyce and Gordon Moore leave Fairchild Semiconductor to found Intel, which creates the first commercially available microprocessor in 1971.

JULY 1970

Xerox Palo Alto Research Center (PARC) opens, eventually developing laser printing, Ethernet, personal computers, graphical user interfaces, and the desktop paradigm, object-oriented programming, ubiquitous computing, a-Si applications, and VLSI for semiconductors.

A TIMELINE OF SILICON VALLEY

January 1971
Don Hoefler brands phrase "Silicon Valley" in a series of articles in the paper Electronic News inspired by friend Ralph Vaerst, a central California entrepreneur.

APRIL 1976
Biochemist Herbert Boyer and venture capitalist Robert Swanson found Genentech, the first major biotech company.

JANUARY 1984
Apple introduces the Macintosh with its graphical user interface, which revolutionizes desktop publishing.

NOVEMBER 1985
Microsoft releases Windows operating system, which comes to dominate the world's personal computer market.

JUNE 1993
Adobe Systems releases Acrobat and the PDF format four years after shipping Photoshop.

SEPTEMBER 1998
Larry Page and Sergey Brin found Google with its search engine based on the PageRank search ranking algorithm.

MARCH 2000
Dot-com bubble reaches its peak before popping over two years. Bubble starts with the release of Marc Andreessen's Mosaic web browser in 1993.

MARCH 2006
Jack Dorsey creates Twitter, then Square three years later.

JULY 2008
Apple's App Store launches alongside the iPhone 3G, generating a marketplace with a massive outreach for app developers. This opportunity creates a wave of mobile software startups in Silicon Valley and San Francisco.

JUNE 2012
Elon Musk introduces Tesla Model S, marking an evolution in automobile engineering.

OCTOBER 2012
Facebook grows to 1 billion users, after originally only being a network for university students.

MARCH 2017
Silicon Valley is global.

4

LONDON

The cosmopolitan capital of the world and dubbed the "Silicon Roundabout," London is home to 300 different languages as well as the highest concentration of higher education in Europe—meaning its sophisticated accent isn't just for show. London is much like New York in terms of high real estate prices, lack of developers, large amounts of beautiful people who could be models instead of entrepreneurs, government incentives, and exposure to industries outside the tech bubble. However, London does supersede New York in terms of global relevance, being home to over 100 of Europe's 500 largest companies. While growth-hindered and culturally still risk-averse, London-based Unicorns who were once a dream are now a reality, with Shazam's $1 billion valuation. As they say in London, "Mum, I made it."

VENTURE ACTIVITY Q4'15 - Q3'16

VC FUNDING	$4.0B
DEALS	694
EXITS	256
5 YEAR YoY FUNDING GROWTH	9.7%
5 YEAR YoY DEAL GROWTH	38.1%

Source: *CB Insights*

Power Breakfast

- Cecconi's
- Regency Café
- Sacred
- The Breakfast Club
- The Wolseley

Coffee Shops to Start Up In

- Curators Coffee Studio
- Ozone
- Shoreditch Grind
- The Attendant
- The Proud Archivist
- Timberyard

Local Brewery

- Brewdog
- Sam Smith's

Offline Tinders

- Egg London
- Electric Brixton
- Fabric
- KOKO
- Ministry of Sound
- Shoreditch House
- Studio 338
- The Book Club

"Must Be At" Local Events

- BBC Proms
- Bonfire Night
- End of Summer Bash
- London Fashion Week
- London Film Festival

Startup District

- Shoreditch

THE TOP

ANGELS

- Barry Smith
- Benjamin Ling
- Christine Tsai
- Jonnie Goodwin
- Neil Hutchinson

VCs

- Accel
- Atomico
- Balderton
- Index
- Lepe
- Octopus
- Piton

VC LAWYER

- Tina Baker

ACCELERATORS

- Bethnal Green Ventures
- Emerge
- EntrepreneurFirst
- Firestartr
- Level 39
- Seedcamp
- Techstars

ENTREPRENEURSHIP EVENTS

- Business Networking London
- Entrepreneurs in London
- Founders Forum
- London Business Starters
- Open Coffee
- Silicon Drinkabout
- Women in Business Conference

ENTREPENEURIAL HEROES

- James Dyson
- Richard Branson

Q4'15 - Q3'16 TOP FUNDING ROUNDS

- Ion Investment Group—$400M
- Deliver—$275M
- Wireless Infrastructure Group—$106.8M
- Ebury—$83M
- Sirin Labs—$72M

PUB **LIFE**

The Silicon Roundabout can quickly become the Silicon Drinkabout. Here are some tips to London pub life.

Not a bar "Pub" is short for "public house," which means pubs aren't just dark, sweaty places where twenty-somethings touch each other. Pubs can range from relaxed old-timers in the country to sixteen-year-olds with fakes yelling about who Chelsea's best player is.

Pub grub While meals are not the norm, pubs do offer a list of heart attack–inducing snacks, including meaty sandwiches, crisps (chips), and "pork scratchings," which may or not be made with itchy pigs and backscratchers.

Drinks It is perfectly fine to ask for a coffee or Coke at an English pub, but fancy cocktails are a no-no, even for the fruitiest of drinkers. Don't ask for a beer either—ask for some bitter (brownish-red ale that's warmer than American beer), lager (clear, light, and generally cold), or a host of other varieties. The go-to size is a pint, while a standard drink is a half-pint, so specify size if you're looking to drink less.

Service If you sit down at a pub expecting someone to serve you, you might as well have justified the 170-year-old oppression of the American spirit up until 1776. Order at the bar, and don't worry too much about tipping. If the bartender helps clean your belligerent uncle's mess, show him your gratitude with a simple "one for yourself," which gives the bartender an excuse to drink more.

London Fintech

In the rising fintech (financial-tech) market, London-based startups are having a huge impact.

GoCardless
Processes direct debit payments
Funds Raised: $11.8M

TransferWise
Low-cost money transfers over borders
Funds Raised: $90.3M

Digital Shadows
Provides enterprise advice about digital footprints
Funds Raised: $8M

Funding Circle
Peer-to-peer platform for lending
Funds Raised: $273.2M

World Remit
Global transfer business
Funds Raised: $140M

Nutmeg
Small investment management services
Funds Raised: $37.3M

DRAGON'S **DEN**

London's version of *Shark Tank* (the concept of both shows originated in Japan in 2001) called *Dragon's Den*, has the same cast of predatory multimillionaires acting like billionaires...except they are no longer aquatic.

WHEN TO PITCH	WHAT TO EXPECT	WHOM TO TRUST
A person should think about pitching to *Dragon's Den* when all venture capitalists have told that person, "kthxbai," when he does not have confidence in his ability to run a business, or when said person already has venture backing but would love some extra press (at the expense of listening to these dragons talk).	During your pitch, the entire room will suddenly freeze for 5–10 seconds, as the camera zooms in on British Mark Cuban's gaping mouth when you tell him you don't want ten thousand dollars for 70% equity. In the background, dramatic music will play. Also, expect cameos from guest dragons like *Lord of the Rings*' Smaug and *Mulan*'s Mushu.	At times, many of these dragons will jump in on a deal: offering varying amounts of equity, money, and pointed remarks directed toward you and each other. Remember that whatever the deal, it will always come with less guidance than advertised.

GOVERNMENT & BUSINESS

It takes a village to raise a child, and a lot more than that to raise a startup. Governments are increasingly becoming more involved with the startup scene, putting time and money into building a startup infrastructure.

The London Co-Investment Fund Uses £25 million from the Mayor of London's Growing Places Fund to coinvest in seed rounds, helping alleviate the lack of VC money in London

Start Up Loans UK Funded by the UK government, it aims to provide mentorship and loans for UK-based startups.

Tech City UK Created in 2010 by UK Prime Minister David Cameron, Tech City UK has a host of programs supporting entrepreneurship and technology, like their Digital Business Academy. Tech City also serves as a connection between entrepreneurs and the government.

TOP 15 PRIVATE COMPANIES BY VALUATION, 2016

Company	Valuation
UBER	$62.5B
ANT FINANCIAL	$60.0B
XIAOMI	$45.0B
DIDI CHUXING	$33.7B
AIRBNB	$30.0B
PALANTIR	$20.0B
SNAPCHAT	$19.3B
LUFAX	$18.5B
MEITUAN-DIANPING	$18.0B
WEWORK	$16.9B
FLIPKART	$15.0B
SPACEX	$12.0B
PINTEREST	$11.0B
DROPBOX	$10.4B
STRIPE	$9.2B

■ Government interaction core to business.
■ Government interaction not core to business.

For a full list of the Unicorns, see page 204.

5

BOSTON and CAMBRIDGE

From the ups and downs of the once booming Route 128, the greater Boston area has performed a successful resurgence, roughly paralleling the career of noted Boston Super Bowl superstar and phone assassin Tom Brady. Now it's all about Kendall Square, where companies are benefiting from world-class talent at the many universities in the area along with the helpful, tight-knit tech community, which hosts everything from Mobile Mondays to the Boston Startups Twitter (@BOSstartups). Though it took far too long for Harvard dropout Mark Zuckerberg to finally start a Facebook office in Boston, having decided to build his company in the Valley instead, the Boston area boasts an impressive healthcare and biotech company record, and is far more practical and sane than the Valley will ever be.

AGGREGATE VENTURE ACTIVITY Q4'15 - Q3'16

VC FUNDING	$6.5B
DEALS	453
EXITS	86
5 YEAR YoY FUNDING GROWTH	13.9%
5 YEAR YoY DEAL GROWTH	6.8%

Source: *CB Insights*

BOSTON DO'S AND DON'TS

✔ **KNOW YOUR SPORTS** Boston takes pride in their sports teams. It's fine if you don't love the Celtics or the Red Sox, but know about them, because if you don't, you'll have absolutely nothing to talk about with any person you meet in Boston. And remember, rumor has it the Boston Massacre began with a publicly worn Yankees hat.

✔ **USE PUBLIC TRANSPORTATION** Parking spaces in Boston are as common as Windows Phones in Silicon Valley. The word "subway" refers to infamous subpar sandwich franchises, whereas the transportation system is called "the T" by any true Bostonian.

THE DON'Ts

✘ **FAKE A BOSTON ACCENT** Locals love their "lobstah" and "chowdah," but the missing r's and elongated a's are an art, with imitators being given the same look as Bill Bellichek's perpetually disgusted resting face.

✘ **BE TOO FRIENDLY** Sure, you want to get to know people, but smiling all the time and asking questions isn't the Boston way. Bostonians would help a blind man cross the street, but wouldn't want to be friends with him afterward.

THE TOP

ANGELS

- Art Papas
- Bill Sahlman
- Chris Keller
- Dharmesh Shah
- Jen Lum
- Jere Doyle
- Joe Caruso
- Lars Albright
- Peter McKay
- TJ Mahony

VCs

- Atlas Venture
- Bain Capital
- Battery
- Bessemer
- CRV
- General Catalyst
- Highland Capital
- Nextview
- North Bridge
- OpenView
- Polaris
- Spark Capital

ACCELERATORS

- Bolt
- Education Program
- HealthBox
- LearnLaunchX
- MassChallenge
- Techstars
- The Capital Network: Accelerated

ENTREPRENEURSHIP EVENTS

- BostInno 50 on Fire
- BUILD Entrepreneur Games
- FutureM
- Greenhorn Connect
- HBS Cyberposium
- Tech Gives Back
- WebInno

BOSTON AT A GLANCE

Power Breakfast
- Henrietta's Table
- Paramount
- Tatte
- The Friendly Toast
- Zaftigs Delicatessen

Coffee Shops to Start Up In
- Pavement Coffeehouse
- Render Coffee
- Thinking Cup

Local Beers
- Batch Whiskey
- Harpoon
- Narragansett
- PBR
- Sam Adams
- Trillium

Best Bars
- Commonwealth
- Gather
- Legal Rooftop
- Meadhall (Kendall Square)
- The Sevens
- Venture Cafe
- WeWork Parties

Startup Districts
- Back Bay
- Cambridge
- Financial/Leather District
- Fort Point (up-and-coming)
- Red Line (Alewife to Dtwn. Crossing)
- Seaport/Innovation District

Boston Networking Tips
- Namedropping shows you're a poser
- Wear Bonobos, sponsor of many of the podcasts you gave up on

"Must Be At" Local Events
- Boston Marathon
- Chowderfest
- St. Patrick's Day

BIG SUCCESS STORIES
- Acquia
- Actifio
- Brightcove
- Care.com
- DataGravity
- Demandware
- DraftKings
- Endeca
- Hubspot
- Kayak
- Kiva
- Quattro
- Rapid7
- Starent
- TripAdvisor
- Trusteer
- VistaPrint
- VMTurbo
- Wayfair
- Zipcar

Q4'15 - Q3'16 TOP FUNDING ROUNDS
- Moderna Therapeutics—$600M
- DraftKings—$153M
- Indigo—$100M
- Ginkgo Bioworks—$100M
- Society of Grownups—$100M

Podcasts Aren't Dead

Most college students are surprised to find out that podcasts still exist—here are some of the "must listen to" Boston Podcasts.

TECH IN BOSTON
Twice a month interviews with local entrepreneurs.

INNOVATION HUB
Features "creative thinkers," interesting people from a wide variety of fields.

#TECHITFWD
Interviews with Boston entrepreneurs, both profit and nonprofit.

THE BOSTON COLLEGE SCENE **BY THE NUMBERS**

School	Fall '15 Acceptance Rate	Undergraduates	Tuition	Entrepreneurs
Harvard University	6.0%	6,694	$45,278	Bill Gates, Mark Zuckerberg
	Clubs: Harvard Ventures, Harvard Innovation Lab			
Massachusetts Institute of Technology	7.9%	4,512	$46,704	Bill Hewlett, Drew Houston
	Clubs: MIT E-Club, Sloan Entrepreneurship and Innovation Club			
Tufts University	17.3%	5,188	$50,604	Pierre Omidyar
	Clubs: Tufts Entrepreneurial Network, Entrepreneurship Immersion Program			
Boston College	33.9%	9,154	$49,324	Nikesh Arora
	Clubs: Edmund H. Shea Jr. Center for Entrepreneurship, Entrepreneurship Society			
Northeastern University	32.3%	13,510	$45,530	Shawn Fanning, Biz Stone
	Clubs: Northeastern Entrepreneurs Club, Entrepreneurs Immersion Program			
Boston University	34.5%	18,017	$48,436	Mickey Drexler
	Clubs: BU Finance and Investment Club, BU Entrepreneurship Club			

Boston's Searching for Stars

Of its nine locations, the Techstars accelerator in Boston has seen some of the biggest successes, along with a huge number of applications. Here's how to get in.

1. **Research.** Make Paul Graham's essays into an audiobook and play it whenever you drive, watch Sam Altman's class while on the can, parse through every tweet of every partner of every accelerator, and please for the love of God, know about every nook and cranny of your industry, and then, maybe, you might be ready for the interview.

2. **Iterations.** Products are only completely and totally finished when your company is, which means iterations on iterations on iterations. Change roles from passionate entrepreneur when making the product to disillusioned college grad in the dot-com bubble when critiquing the product.

3. **Upfront.** You aren't pitching to a venture capitalist, who only wants to hear about your 95% share of a $16 trillion market in ephemeral photo applications. Accelerators are meant to help, so evaluate your company objectively, and be optimistic but honest.

BEIJING and SHANGHAI

Beijing and Shanghai as cities differ in terms of dialect and culture, food and pollution (only Beijing has a Twitter feed dedicated to reporting the pollution each day), but both represent the meteoric rise of the China startup ecosystem, bursting onto the global scene like firecrackers and noise complaints during Chinese New Year.

Scale is simultaneously a problem and not a problem for the Chinese—China's colossal 1.4 billion population is a huge consumer base (with an insane amount of cheap manufacturing) but companies often fall prey to an inability to expand past China. A large majority of startups are focused on the internal market. And if you thought copycat companies were a problem in America, it's on a whole 'nother level out in the East, from fake Apple stores to coffee shops called Sunbucks. China will need to catch up to the rest of the world in terms of innovative ideas, but they boast a hardworking culture and an unparalleled national consumer base.

BEIJING VENTURE ACTIVITY
Q4'15 - Q3'16

VC FUNDING	$20.5B
DEALS	213
EXITS	44
5 YEAR YoY FUNDING GROWTH	90.0%
5 YEAR YoY DEAL GROWTH	28.4%

Source: *CB Insights*

SHANGHAI VENTURE ACTIVITY
Q4'15 - Q3'16

VC FUNDING	$9.5B
DEALS	137
EXITS	30
5 YEAR YoY FUNDING GROWTH	118.7%
5 YEAR YoY DEAL GROWTH	39.4%

Source: *CB Insights*

C2C: COPY TO CHINA

China's Great Wall has become the Great Firewall, insulating Chinese consumers from foreign companies. This, in turn, has produced a host of extremely profitable Chinese companies with similar functions to the American tech giants, creating Chinese billionaires in the process.

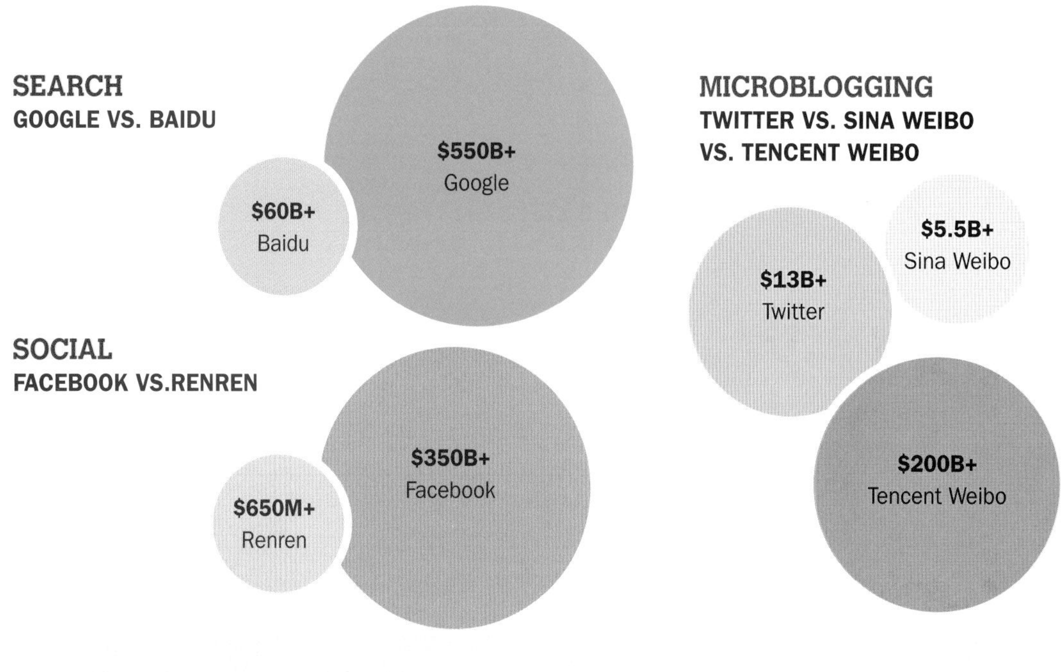

Beijing Networking Tips

- Have Chinese versions of your promotional materials.
- Be respectful of government issues and be more PC than you are in the States.
- Good business relationships require time.

BEIJING AT A GLANCE

Power Breakfast
- Chef Too
- Grandma's Kitchen
- Le Cafe Ingrosso
- The Courtyard

Coffee Shops to Start Up In
- 3W
- Beta
- Cheku

Local Beers
- Great Leap
- Malty Dog
- Panda Brew
- Slow Boat
- The First Immortal DIPA
- Tsingtao

Offline Tinders
- Babyface
- Elements
- Fang Bar
- Mix
- Propaganda
- Vics
- World of Suzie Wong

"Must Be At" Local Events
- Chinese New Year
- Mid-Autumn Festival

Startup District
- Haidan
- Wanjing

THE TOP

ANGELS
- Binbin Zhang
- ChengCheng Liu
- Chris Evdemon
- James Tan
- Lucas Wang
- Phil Morle
- William Bao Bean

VCs
- Banyan Capital
- DCM
- DT Capital Partners
- Fortune Venture Capital
- GSR
- Hopu
- IDG Capital Partners
- Innovation Works
- Sequoia Capital China
- Tencent
- Zhenfund
- ZZ Capital

TOP ACCELERATORS
- 36Kr
- FutureWorks
- Microsoft Accelerator Beijing
- Seedit
- UrWork
- XLab

ENTREPRENEURSHIP EVENTS
- EmTech Beijing
- GMIC Beijing
- Startup Weekend China
- TechCrunch Beijing

BIG SUCCESS STORIES
- Baidu
- JD.com
- Lenovo
- TAL Education
- Wanda
- Xiaomi Technology

Q4'15 - Q3'16 TOP FUNDING ROUNDS
- Meituan—$3.3B
- LeSports—$1.2B
- JD Finance—$1B
- SECA—$1B
- Lianjia—$926M

SHANGHAI AT A GLANCE

Power Breakfast
Jean Georges
Jing'An Shangri-La Hotel

Coffee Shops to Start Up In
Cambio Marketplace
ICCafe
IPO Club
JiuCengGeCafe
Punk Coffee
SeeSaw Cafe
Sumerian
WeeCoffee

Local Beers
Boxing Cat
Dr. Beer
Imperial Stout
King Louie

Offline Tinders
390
Hollywood
Kai Bar
Le Baron
Lola
Monkey Champagne
TZ
Unico

"Must Be At" Local Events
Chinese New Year
Dragon Boat Festival
Lantern Festival

Startup Neighborhood
Caohejiang Tech Park
Wujiaochang
Zhangjiang Research Inovation Park

THE TOP

ANGELS
Cyril Ebersweiler
James Haft
Jonathan Tam
Wei Guo
William Bao Bean

VCs
Greenwood
GSR
IDG
Lightspeed
NEA
QimingVenture Partners
Sequoia Capital China
Tencent

TOP ACCELERATORS
Chinaccelerator
FeiMaLv
Innospace
Innovation Camp
iStart Ventures
SEEDiT
Sinovations
Shanghai Cloud Valley
Shanghai Technology EFG
Shanghai Valley
SLP Shanghai
Startup Commune
SuHeHui
xNode

ENTREPRENEURSHIP EVENTS
GSMA Innovation City
GTI Summit
Mobile World Congress Shanghai
TechCrunch Shanghai

Q4'15 - Q3'16 TOP FUNDING ROUNDS
Ele.me—$1.25B
Ctrip—$1B
Lufax—$924M
BabyTree—$450M
Dada—$300M
Daojia—$300M

BEIJING'S **INNOVATIVE SPACES**

36KR

- Leading online tech news website in China as well as an accelerator.
- Began as a TechCrunch translator site, but eventually rebranded and started publishing original content.
- Companies in the accelerator program are often given heavy media exposure.

BINGGO SCHOOL

- Features a startup café, accelerator, and angel fund.
- Connected to Virtue Innovalley, an entrepreneurship mentoring program out of Tsinghua University.
- Partnered with 1776, a global incubator and venture fund, and Microsoft Ventures to host the 2015 Challenge Cup, a hackathon-like event that is designed to help startup founders connect with influencers and partners.

Z-INNOWAY

- Z-innoway, which stands for Zhongguancun Innovation Way, opened in June 2014, and is a 450,000-square-foot project in Beijing dedicated to entrepreneurship. The site is home to startup cafés, accelerators, coworking spaces, and membership-based entrepreneurship clubs. It's like Facebook's campus...on steroids. Going back to "Phrases to Avoid," we're not saying this campus is bald and sterile.

3 W

- Originally crowdfunded by influential angels like Neil Shen of Sequoia China.
- Has an angel fund that typically invests about $80,000 in select companies in the accelerator.
- Startup café is located on the first and second floors, and the Next Big Accelerator is on the third.
- Was visited by Premier Li Keqiang, an important Chinese politician, in May 2015, revealing the growing interest of the Chinese government in entrepreneurship

AVERAGE JACK **NO LONGER**

From English teacher to the second richest man in China, Jack Ma hasn't taken the conventional path to being an entrepreneur, but he has been merely five feet tall and driven—talk about scrappy. From dressing up and performing as a pop star with a white wig and black leather jacket in front of his employees to starting an original website in China when most Chinese companies were and continue to be copycats, Jack Ma is part rags-to-riches, part eccentric billionaire, part psychotic war general, part philanthropist, and complete genius.

1964 Jack Ma was born Ma Yun (family name Ma) in Hangzhou. His parents worked as professional pingtan performers, a traditional Chinese art combining humor, story, and music. His mother was also a factory worker.

1970s Ma worked as a free tour guide for English-speaking foreigners to improve his English, eventually becoming pen pals with a girl who gave him the name "Jack." And you thought *The Notebook* was romantic.

1988 After failing the college entrance exam twice, Ma attended and graduated from the Hangzhou Teachers Institute, earning a degree in English. After graduation, Ma worked as an English teacher at Hangzhou Electronic and Engineering Institute, a lesser-known institution, later founding a translation agency.

1995 Ma traveled to the US for the first time, going to Malibu to collect a debt from an American businessman. This American refused to pay the debt, and Ma was instead forced to accompany the businessman on a trip to Vegas. Looking for a way out, Ma played the slot machines, winning $600, enough to buy him a plane ticket to Seattle. In Seattle he discovered the computer and the Internet, typing the word "beer," which produced a host of results, and then "beer China," which produced nothing. This spawned Ma's first company, a Chinese version of the Yellow Pages, which failed.

1998 While working at the Ministry of Foreign Trade and Economic Cooperation, Ma took Yahoo!'s Taiwan-born cofounder Jerry Yang on a tour of the Great Wall.

1999 Ma gathered 17 of his friends, raised $60,000, and founded Alibaba in his apartment. By the end of the year, Ma had raised venture capital money from Goldman Sachs and SoftBank, a Japanese telecom company.

2001 In December, Alibaba surpassed 1 million registered users.

2003 Alibaba launched Taobao.com, an online shopping site similar to eBay and Amazon. At the time, eBay controlled much of the Chinese market share, and Ma announced he was "going to war" with the company.

2005 Alibaba formed a "strategic partnership" with Yahoo!, buying a 40% stake in the company for $1 billion. Alibaba later took control over Yahoo! China's operations.

2007 Alibaba IPO'd in Hong Kong and lists on the Hong Kong Stock Exchange

2009 In the celebration of the ten-year anniversary of the company, Ma dressed up in a long, white hair wig and a black leather suit and sang *The Lion King*'s "Can You Feel the Love Tonight?" in front of over 16,000 employees. Eccentric genius takes many forms.

2013 Jack Ma stepped down as CEO of Alibaba, retaining his chairman role.

2014 Alibaba had the largest ever IPO in US history, raising $25 billion.

2016 Alibaba's market value is over $230 billion.

THE **ALIBABA** EMPIRE

Alibaba is more than something that exists solely in the confines of *Aladdin*. Alibaba had the biggest IPO in the United States when it went public in 2014 on NASDAQ. What started as an e-commerce company founded by Jack Ma in 1999 eventually grew to become a monster business with a suite of services in electronic payments, online shopping, cloud computing, and more. **Welcome to the Baba Dynasty.**

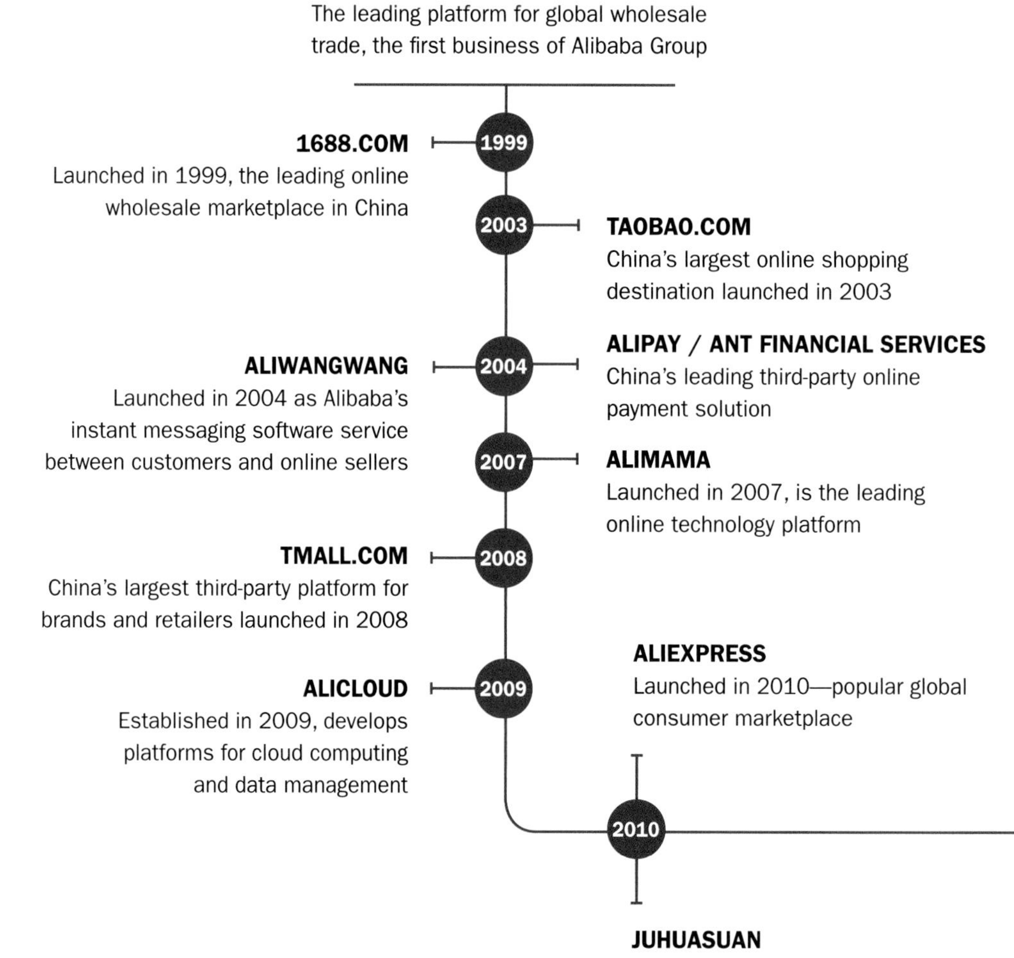

The Expat's Experience

The strong connection between the developed economies of China and America has given rise to extended travel between the two countries. Expatriates, or expats, are people who live outside their native country—here's an expat's guide to the differences between the US and China.

Freedom. Information in China is much harder to access than in the United States. The Great Firewall of China rains flaming arrows on anything deemed by the Chinese government to be unfit for web consumption, blocking everything from Gmail to Bloomberg, resembling the parental controls you encountered when first attempting to watch explicit videos at home. *VPNs, or virtual private networks, will help you circumvent these restrictions.*

Competition. Competition is fierce in both countries, but in China it is particularly stiff. *Copycat websites are more socially acceptable*, meaning the moment your brilliant "Tinder for wireless routers" app is created, it will be stolen, stripped bare, put in Google translate, and no one will know which came first.

Résumé. In China, having US startup experience can be good, but be wary of swaggering in and acting like you know everything better—*it's a different world in China, and all your valuable knowledge and accumulated network need updating*. In the US, having China startup experience can open up many job opportunities, but can also pigeonhole you as the "China guy/gal." Not only will job offers be geared toward that expertise, but coworkers will now demand you to tell them whether or not Panda Express is authentic, though some of us are enjoying the "ignorance is bliss" of the honey walnut shrimp.

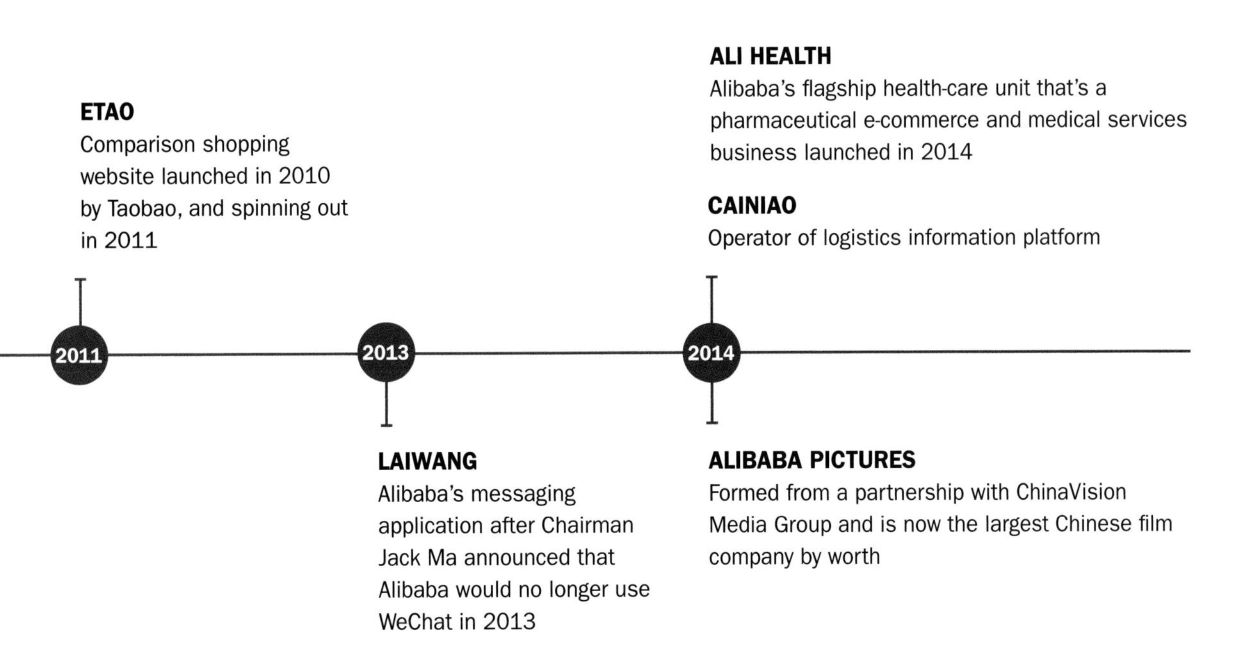

CHINA'S OTHER **POWERHOUSES**

HANGZHOU POP. 6,532,000

Renowned for its natural beauty, Hangzhou has tourists aplenty, but has also been lauded as a very business-friendly city. Home to Alibaba.

VENTURE ACTIVITY Q4'15 - Q3'16

VC FUNDING: $6.1B	DEALS: 27	EXITS: 4
5 YEAR YoY FUNDING GROWTH 224.7%		
5 YEAR YoY DEAL GROWTH 31.0%		

Source: *CB Insights*

SHENZHEN POP. 12,337,000

Shenzhen is home to the headquarters of numerous tech companies (like Tencent) and its own stock exchange, while also serving as a manufacturing hub with a focus on more advanced electronics.

VENTURE ACTIVITY Q4'15 - Q3'16

VC FUNDING: $1.3B	DEALS: 42	EXITS: 5
5 YEAR YoY FUNDING GROWTH 230.9%		
5 YEAR YoY DEAL GROWTH 47.6%		

Source: *CB Insights*

While Beijing (with a population of over 18 million) and Shanghai (with a population of 23 million) lead China in terms of VC funding and deals, China's other powerhouse cities are upcoming leaders in the growth economy. China has 160 cities with a population over 1 million, each with an ambition that matches it's size and growth.

SUZHOU POP. 4,367,000

Called the "Venice of the East," Suzhou is a major trade center—especially for silk. The city is also known for its beautiful gardens and cultural history and is where famed architect I. M. Pei calls home.

VENTURE ACTIVITY Q4'15 - Q3'16

VC FUNDING: $203M	DEALS: 8	EXITS: 2
5 YEAR YoY FUNDING GROWTH		NA%
5 YEAR YoY DEAL GROWTH		NA%

Source: *CB Insights*

GUANGZHOU POP. 12,385,000

A manufacturing juggernaut and southern China's leading industrial and commercial center, Guangzhou has factories that export huge amounts of plastics, electronics, and clothes worldwide.

VENTURE ACTIVITY Q4'15 - Q3'16

VC FUNDING: $283.7M	DEALS: 11	EXITS: 4
5 YEAR YoY FUNDING GROWTH		78.3%
5 YEAR YoY DEAL GROWTH		61.5%

Source: *CB Insights*

CHENGDU POP. 7,815,000

The biggest urban area not in East China has a more relaxed lifestyle, and serves as the anchor for West China as well as a host of big international companies. Home of giant pandas and birthplace of Li Jiang.

VENTURE ACTIVITY Q4'15 - Q3'16

VC FUNDING: $150.7M	DEALS: 4	EXITS: 1
5 YEAR YoY FUNDING GROWTH		72.0%
5 YEAR YoY DEAL GROWTH		32.0%

Source: *CB Insights*

HONG KONG POP. 7,431,000

Once part of the British Empire, Hong Kong is now its own autonomous entity under the Chinese government, a truly global financial center that rivals London and New York. See page 168 for more info.

VENTURE ACTIVITY Q4'15 - Q3'16

VC FUNDING: $625.0	DEALS: 74	EXITS: 34
5 YEAR YoY FUNDING GROWTH		0%
5 YEAR YoY DEAL GROWTH		35.8%

Source: *CB Insights*

WUHAN POP. 10,256,000

Bordering the Yangtze River, Wuhan has a little bit of everything, known as a transportation hub but also anchoring much of the business and education of Central China. Hometown of tennis star Li Na.

VENTURE ACTIVITY Q4'15 - Q3'16

VC FUNDING: $15M	DEALS: 2	EXITS: 2
5 YEAR YoY FUNDING GROWTH		N/A
5 YEAR YoY DEAL GROWTH		N/A

Source: *CB Insights*

NANJING POP. 6,723,000

Rich in Chinese history and culture, Nanjing is an education center in China and is far less urban than Shanghai or Beijing.

VENTURE ACTIVITY Q4'15 - Q3'16

VC FUNDING: $71.9M	DEALS: 9	EXITS: 0
5 YEAR YoY FUNDING GROWTH		NA%
5 YEAR YoY DEAL GROWTH		NA%

Source: *CB Insights*

The Apple Tree

For its millions of employees and thousands of companies, Silicon Valley is home to a group of deeply intertwined tech giants who have benefited from the mentorship and example of those ahead of them. Here is Steve Jobs's coaching tree.

Known as simply "Coach" in the Valley, Bill Campbell used to coach Columbia University's football team, but moved out West to work for Apple under Jobs and eventually became the CEO of Intuit. His nickname comes from the large number of tech entrepreneurs he has carefully mentored to success.

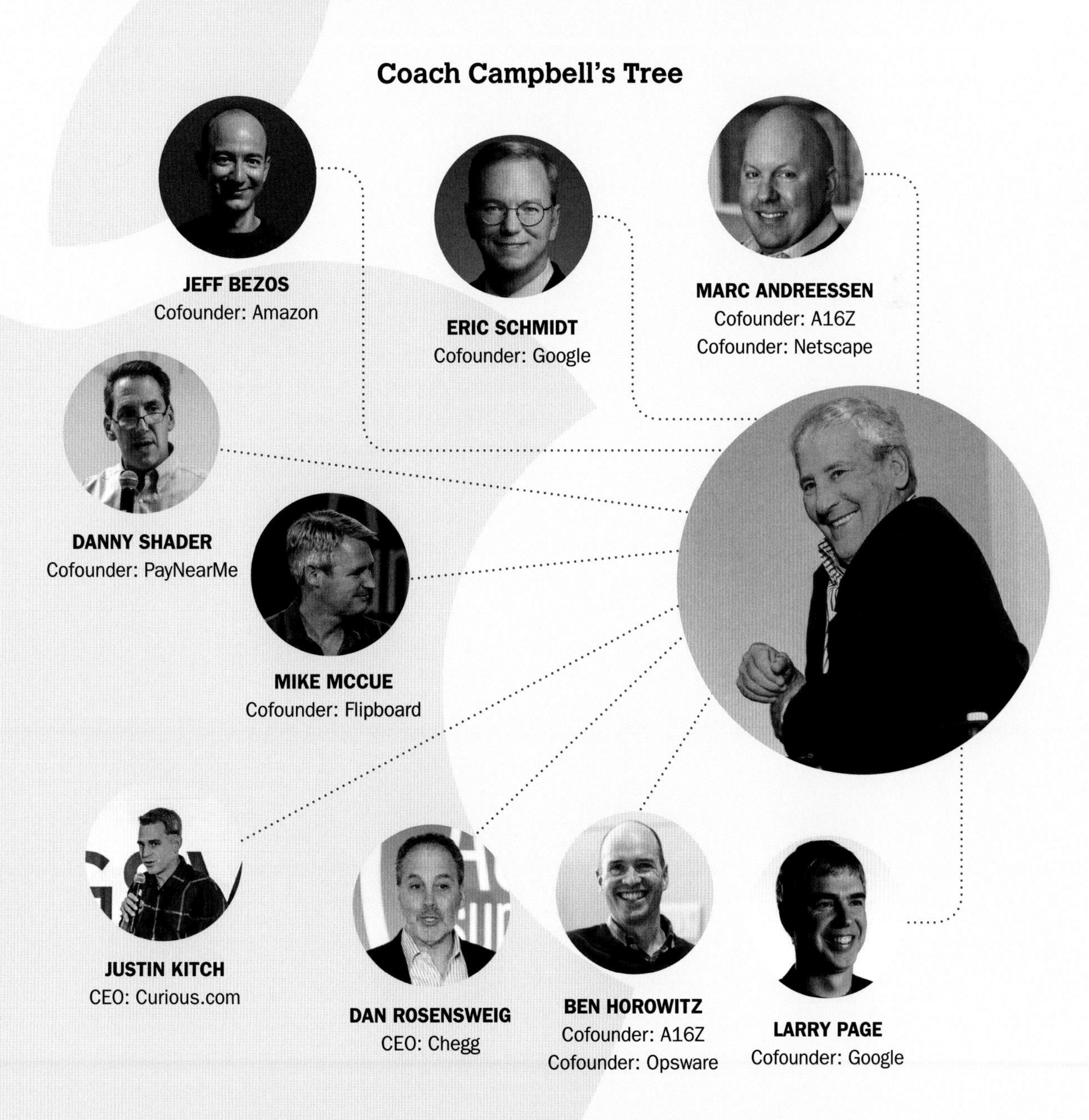

8

CHICAGO

Before the region was known as "Silicon Prairie," VCs called Chicago "the flyover city" as they jetted from Silicon Valley to Boston looking for startups to invest in. As no industry dominates the city, startups in Chicago are diverse. And with colleges like Northwestern, Urbana-Champaign (UIUC), and University of Chicago so close by, the talent is all around. Treated like an ex by the founders of Netscape, PayPal, YouTube, and Yelp—who all ventured to the Valley to begin their companies—Chicago finally found a friend in Groupon, where the fingerprints of cofounders Eric Lefkofsky and Brad Keywell are still left all over the Chicago tech scene. With its affordable prices and insistence on no ketchup on hotdogs, the Chicago startup scene hopes to embody the values of its largest incubator, 1871, named after not the Great Chicago Fire, but the innovation and teamwork that came after it.

VENTURE ACTIVITY Q4'15 - Q3'16

VC FUNDING	$1.8B
DEALS	275
EXITS	100
5 YEAR YoY FUNDING GROWTH	14.6%
5 YEAR YoY DEAL GROWTH	10.8%

Source: *CB Insights*

CHICAGO **AT A GLANCE**

Power Breakfast
- Chicago Cut
- Lou Mitchell's
- Pierrot Gourmet
- The Bagel
- Yolk

Coffee Shops to Start Up In
- Goddess & the Baker
- Intelligentsia
- La Colombe
- Sip Coffee House

Best Bars
- Big Star
- Lotties
- Lux
- Parlor
- RL
- The Aviary

Local Beer
- 312
- Forbidden Root
- Goose Island
- Old Style

Go-to Shot
- Fireball
- Malört

Offline Tinders
- Delilahs
- Fado
- Original Mother's
- Richards Bar
- Soho H.
- The Hangge Uppe
- Underground

"Must Be At" Local Events
- Air and Water Show
- Art Expo
- Blackhawks/Cubs/Bulls Game
- Chicago Gournet
- Lollapalooza
- Old Town Art Fair
- Navy Pier Fireworks
- Second City
- Taste of Chicago
- TJ & Dave at iO

THE TOP

ANGELS
- David Cohen
- Deborah Quazzo
- Harper Reed
- Howard Tullman
- Kevin Willer
- Jason Fried

VCs
- Apex Venture Partners
- GSV Acceleration
- GSV Ventures
- Hyde Park Angels
- LightBank
- Pritzker Group
- True North
- Valor Equity

TOP ACCELERATORS
- 1871
- BLUE1647
- Catapult
- HealthBox
- Impact Engine
- Matter
- TechNexus
- Techstars
- U of Chicago New Venture Challenge
- University Technology Park at IIT

ENTREPRENEURSHIP EVENTS
- 1871 TGIFs
- Blogcademy
- BMA
- Chicago Startup Weekend
- Chicago Venture Summit
- Techweek Chicago

Q4'15 - Q3'16 TOP FUNDING ROUNDS
- Avant—$325M
- DuPage Medical Group—$250M
- The Onion—$200M
- SMS Assist—$150M
- Vibes Media—$45M
- Uptake—$45M

THE PAYPAL **MAFIA**

Chicago was once home to one of the most powerful mob groups of all time—the PayPal Mafia, many of whom once called UIUC their home. These individuals have had big roles in quite a few legendary companies.

TEAM MEMBER	PAYPAL POSITION	AFTER PAYPAL
Elon Musk	Cofounder, Director	Founder and CEO, SpaceX Cofounder and CEO, Tesla
Peter Thiel	Cofounder, CEO	Managing Partner, Founders Fund President, Clarium Capital Board Director, Facebook
Max Levchin	Cofounder, CTO I	Chairman, Yelp; CEO, Affirm Executive Chairman, Glow Founder, Slide (acquired by Google)
Reid Hoffman	Executive VP	Cofounder and Executive Chairman, LinkedIn Partner, Greylock
Keith Rabois	Executive VP, Business Development	Partner, Khosla Ventures COO, Square Investor, YouTube, Geni, LinkedIn, Yelp
Roelof Botha	CFO	Managing Partner, Sequoia Capital
Jeremy Stoppelman	VP Engineering I	Cofounder and CEO, Yelp
Russel Simmons	Engineer I	Cofounder, Yelp
Steve Chen	Engineer I	Cofounder, YouTube
Jawed Karim	Engineer I	Cofounder, YouTube
Chad Hurley	Designer	Cofounder, YouTube
David Sacks	COO	CEO, Zenefits Cofounder and CEO, Yammer Chairman, Geni
Dave McClure	Director Marketing	Advisor, Kiva Founding Partner, 500 Startups
Premal Shah	Product Manager	President, Kiva

Note: UIUC grads are marked with UIUC logo. I

Navigating One of the Best Music Festivals in The US

1 **Cell Phones.** Cell phones present a confounding conundrum for the average Lollapaloozer. On the one hand, they can be used to take pictures and videos for your 160-second snap story documenting the day's antics, but on the other, will have unreliable service and power in the area. As bulky as they are, walkie talkies or messenger hawks could be used as alternatives.

2 **Transportation.** On the one hand, there's the option of driving and parking. This method is sobriety inducing and cost ineffective. The second is public transportation—it can make for a good ride with a couple hundred other drunken festival-goers, but if one person throws up, it's game over. We like cycling—plenty of places to park and the winds of the Windy City helping erase the ringing in your ears after Metallica.

3 **Music.** Choose one stage and stick with it—movement means leaving your friend who is looking for the port-a-potty behind, as well as functionally watching a live performance thousands of feet away on a TV.

WHEN: Last Weekend of July

WHERE: Grant Park, Chicago

CASE STUDY: CHICAGO 1871

The leading incubator/accelerator in Chicago, 1871 houses a community of designers, engineers, and entrepreneurs who benefit from its mentorship and programs. Unique programs at 1871 include WiSTEM, a program for woman entrepreneurs; Bunker Labs, an environment for veteran entrepreneurs; DV X Labs, an EdTech incubator; and the Chicago College Startup Competition, in which ten college businesses receive grants from the organization.

ENTREPRENEURSHIP IN ACTION

Led by Founding President Josephine Lee, Northwestern was able to create its own entrepreneurship club, EPIC (Entrepreneurs Pioneers Innovators Creators). We interviewed the past president of EPIC, Suzee Han.

What does EPIC do? Before EPIC, there was a more fragmented entrepreneurship scene on campus. The past president rebranded and re-created the club with the help of advisors like [GSV Vice President] Li Jiang. I helped Jo create an entrepreneurship development program that spanned nine weeks. It was student run and we brought in professors and CEOs to teach various aspects of how to build a startup. We also launched our first hackathon, had a startup career fair, and had a large business competition that grew to such an enormous size that we couldn't even imagine.

Who was the biggest speaker you brought in? With the help of the Farley Center, we brought in Peter Thiel. It was extremely exciting for us because he's one of the top investors in Silicon Valley, was an early investor in Facebook (not to mention a founder of PayPal and Don of the PayPal Mafia). It was a great event. The event was coincidentally the first Monday of finals week, so people finished their finals early just to attend!

What was the club's vision? Northwestern is a very pre-professional campus. There are a lot of students who strive to be consultants, professors, or bankers—entrepreneurship is growing but it's not quite there yet. As such, we just wanted to create a more entrepreneurial campus because the entrepreneurial mind-set is important no matter what path you take. Our mission was to build an entrepreneurial community for everyone at Northwestern.

9

BERLIN

When the Berlin Wall came down, it marked the end of Soviet domination in Germany and the invasion of partying hipsters in Berlin, Angela Merkel being one of them. And despite critics describing the Berlin startup scene as all hype with no real jobs, as well as the usual problems of scale, it has prospered. With talented young people from all over the world in a vibrant and affordable city that is incredibly welcoming to newcomers, Berlin has overcome the controversies that were once associated with Germany (namely Kim Dotcom, the founder of file sharing website Megaupload, and Fabian Thylmann, once owner of an online porn empire and responsible for billions of acts of self-pleasure) and created a tight-knit startup community with a big creative influence. General Assembly even has a class titled, "Introduction to the Berlin Startup Scene." Speaking German is not necessary, but having a SoundCloud profile with "wet" electronic music is a must.

AGGREGATE BERLIN VENTURE ACTIVITY Q4'15 - Q3'16

VC FUNDING	$1.0B
DEALS	225
EXITS	41
5 YEAR YoY FUNDING GROWTH	32.1%
5 YEAR YoY DEAL GROWTH	22.4%

Source: *CB Insights*

BERLIN **AT A GLANCE**

Power Breakfast
- St. Oberhalz
- The Digital Eatery

Coffee Shops to Start Up In
- Godshot
- Grosz
- Oslo Kaffebar

Best Bar
- Bar 1000

Local Drinks
- Bionade
- Club Mate (with vodka)
- Voelkl Zisch

Offline Tinders
- Berghain & Panorama Bar
- Sisyphos

Startup District
- Kreuzberg
- Mitte

"Must Be At" Local Events
- Berlin Music Festival
- Christmas Markets
- Oktoberfest

THE TOP

ANGELS
- Aleksandr Dresen
- Christoph Marie
- Fred Destin
- Klaus Hommels
- Phillip Moehring

VCs
- EarlyBird VC
- German Startups Grp.
- Lakestar
- Paua Ventures
- Rocket Internet

ACCELERATORS
- Axel Springer Plug&Play
- Berlin Startup Academy
- Hardware.co
- hub:raum
- Microsoft Ventures Berlin
- Startupbootcamp
- Youisnow

ENTREPRENEURSHIP EVENTS
- Berlin Talent Summit
- Build IT Berlin
- Charité Entrepreneurship Summit
- Startup Camp Berlin
- Startup Europe Summit

Q4'15 - Q3'16 TOP FUNDING ROUNDS
- AUTO1 Group—$117.6M
- SoundCloud—$70M
- Azubu—$59M
- GetYourGuide—$50M
- Spotcap—$50M
- Younicos—$50M

Need to Know

Beginning in Berlin in 2007, Rocket Internet bills itself as a startup accelerator and company builder, but more often resembles a copycat company. They traditionally find successful websites in the United States, clone them toward an international market, and sell them back to the original. The Samwer brothers and their model have been wildly successful in terms of raising money, but Rocket Internet has given Berlin a copycat reputation.

THE FOUNDERS

Alexander, the youngest, has an MBA from Harvard and acts as a venture capitalist. Oliver, the middle brother, serves as the CEO of Rocket Internet, with a business degree from WHU, a well-known German business school. Marc, the eldest, has a law degree from the University of Cologne, is uninvolved with the day-to-day activities of the company, and has previously worked extensively with Groupon.

Amid criticism of being idea thieves and innovation killers, the Samwer brothers have argued that while some are better at generating ideas, they are better at executing them. In a rare interview in Berlin, Oliver called good inventors "Einsteins," but compared him and his brothers to Bob the Builder, executors of inventions. Because who would rather be a highly popular cartoon character than a man with the highest IQ?

Rocket Internet had an $8.2 billion IPO. Since then, however, the stock price of the company has declined, leading to questions about its future.

Investment: ZALANDO

Founded by the Samwer brothers as an eBay copycat, and acquired by eBay for $43M in 1999.

Investment: CITYDEAL

In 2010, the Samwer brothers scaled and invested in CityDeal, a knockoff of Groupon that became the top deal site in thirteen countries within five months. Groupon acquired CityDeal, appointing Marc as an executive of international operations. Marc later stepped down from the role amid controversy about a toxic work culture.

Investment: CRAZY FROG

The annoying frog ringtone was marketed by Samwer company Jamba, which was acquired by Verisign for $273M.

Berlin-based company SoundCloud offers free streaming of a wide variety of musicians. Electronic remixes are most popular on the site.

5 Must Follows On SoundCloud

1. **ODESZA** *395K Followers*
Electronic music duo that rose to fame with debut album *Summer's Gone*.

2. **ABSENCE** *49K Followers*
Underground DJ with some remixes of artists ranging from Lana del Rey to Wiz Khalifa.

3. **FLYING LOTUS** *5.5M Followers*
DJ and rapper based out of Los Angeles has very experimental taste.

4. **SOUND PELLEGRINO** *23K Followers*
Paris-based label focusing on club music.

5. **#UNKNOWN** *24K Followers*
London-based account releasing house music with no artist information to dissuade the public from making superficial assumptions by the artist. Mysterious.

THE STARTUP LINE OF BERLIN

The U8 Line not only allows people to travel throughout Berlin, but also stops at many entrepreneurial hotspots. *Originally published on setting.io.*

Voltastraße

Seed VC firm WestTech Ventures built up its incubator Project Flying Elephant in the former supermarket chain Kaisers (previously used by coworking space Supermarkt).

Bernauerstraße

Factory: referred to the new startup campus, where some of Berlin's most prominent startups like SoundCloud, GoButler, and several others call the former brewery their home base.

Rosenthalerplatz

St. Oberholz: the preferred hotspot for Mac users and digital nomads, where WiFi is free, and eavesdropping on startup conversations is included. Mein Haus am See and Neue Odessa Bar on Torstraße are the places where you find Mitte startup employees sipping their cocktails.

Weinmeisterstraße

Hackescher Markt has become somewhat a magnet for ad-tech companies—one of the pioneers being Hitfox, whose office is located on Rosa-Luxemburg Straße. Across the station, startup supportive corporate SAP has its Berlin office.

Alexanderplatz

A few startups have settled around here, including Wunderlist, MeinFernbus, and Wooga. A bit farther south, located on Museumsinsel, is the newly opened German Tech Entrepreneurship Center (GTEC), which has become home to the Berlin Startup Academy and US accelerator Techstars.

Jannowitzbrücke

Immobilienscout24 and its accelerator program YOU IS NOW are just a few blocks away from Jannowitzbrücke.

Moritzplatz

Betahaus is Berlin's most known coworking space and popular starting point for newbies to the city's startup scene.

Kottbusser Tor

GSG-Hof on Adalbertstraße is where startup Locafax rented its office. The offices of eDarling, EyeEm and Blinkist are also just a stone's throw away.

Schönleinstraße

Several startups have found their base along Paul-Lincke Ufer, including SumUp, Mymuesli, and Amorelie. Close by, Umspannwerk hosts companies like shopkick and fitengo. Kreuzberg's Graefekiez is also known as the "Bitcoinkiez," where you can use the virtual currency to pay for your coffee, make photocopies, or pick up a vinyl record.

BERLIN SUCCESS STORIES

ResearchGate
Rocket Internet
SoundCloud
Zalando

ENTREPRENEURIAL HEROES

Alex Ljung
Ilja Madisch
Oliver Samwer
Voltastraße

10

Tel Aviv ain't no Jerusalem—while Jerusalem may be the religious capital of Israel, Tel Aviv is the tech capital. The young beach town is full of young professionals who come from mandatory military training and has the higest density of startups in the world. Startups here avoid the "Snaptrap" or "Instafad" of focusing on social networking or other saturated, productivity-wasting industries—Israel's rocky relationship with its neighbors leads to a focus on more impactful industries, like sustainability and security. And while Americans may have gotten used to the overly fake startup networking events, people are far more blunt and direct in Tel Aviv. With a history of innovation in Israel along with the highly skilled workforce in Tel Aviv, you'll know why citizens of Tel Aviv believe anything is possible.

VENTURE ACTIVITY Q4'15 - Q3'16

VC FUNDING	$983.9M
DEALS	143
EXITS	22
5 YEAR YoY FUNDING GROWTH	60.5%
5 YEAR YoY DEAL GROWTH	44.1%

Source: *CB Insights*

THE TOP

ANGELS

- Daniel Recanati
- Eddy Shalev
- Eilon Tirosh
- Gigi Levy
- Guy Gamzu
- Moshe Lichtman

VCs

- 83North
- Canaan Partners
- Carmel Ventures
- Israel Seed Partners
- StageOne Ventures
- Magma Venture Partners
- Vertex Ventures

ACCELERATORS

- 8200 EISP
- Citi Accelerator
- DreamIT Ventures Israel
- Hamadgera
- IDC Elevator
- Junction
- Microsoft Azure Accelerator
- Startup East
- TheHive by Gvahim
- WMN

ENTREPRENEURSHIP EVENTS

- 4YFN
- DLD Conference
- Israel Dealmaker's Summit
- Israel Mobile Summit
- Israel's Growth Hacking Summit
- TechFest
- The Journey
- WE Summit

Q4'15 - Q3'16 TOP FUNDING ROUNDS

- Gett—$300M
- ForeScout Technologies—$76M
- Fiverr—$60M
- Cyberreason—$59M
- Forter—$32M

Israel's Top Accelerator: ISRAEL DEFENSE FORCES

Military service is mandatory for all Israelis, meaning every entrepreneur in the making has been through service, benefiting from the discipline, network, and training the military provides.

DAN TOLKOWSKY
Cofounder of the first Israeli-based venture capital fund, the Athena Fund
IDF: Commander of the Israeli Air Force

BARAK EILAM
CEO of NICE Systems, big Israeli company focused on data security and surveillance
IDF: Member of Unit 8200, an elite Israeli intelligence organization

CHEMI PERES
Cofounder of Pitango, Israel's leading venture capital firm with $1.6 billion under management
IDF: Pilot in the Israeli Air Force for ten years

URI LEVINE
Cofounder of Waze, mapping company that was sold to Google for $1 billion
IDF: Developer, served five years as opposed to the required three

TEL AVIV AT A GLANCE

Power Breakfast
- Benedict
- Nechama VaHetzi
- Topolopompo

Coffee Shops to Start Up In
- Aroma
- The Streets

Best Bar
- Nanuchka
- Otto
- Rothschild 12
- Rubi
- Social Club

Local Beer
- Alexander
- Goldstar
- Macabi

Best Quick Food
- Frishman Falafel (falafel)
- Tony Vespa (pizza)

"Must be At" Local Events
- Lila Levan
- Nachlat Binyamin Fair
- Tel Aviv Pride Parade

The Wantrepreneur

Citizens of Tel Aviv are renowned for their BS detector—here's who you shouldn't be when networking in the city.

The Constant Marketer Getting your startup idea out there is great, but talking about your startup when it's completely unrelated is a big no-no. Ex. Q: Where are you from? A: My startup is actually transforming iPhones into Sesame Street virtual reality holograms.

The Name Dropper The most common and dreadful sin among young entrepreneurs, and the sure sign of a wanna-be. So your cousin's girlfriend's best friend's aunt twice removed met Steve Jobs in an elevator? Cool story. The true masters of networking never mention who they know—they talk about what they've done.

The VC Hunter Everyone's met that one entrepreneur at a networking event who takes a look at the company on your name tag, smirks, and goes elsewhere. Don't be that guy. The real value of networking is sharing ideas with people and forming connections with them—only chasing company names will get you nowhere.

The map shows the distribution of all these tech establishments throughout Israel.

Source: Mapped in Israel.

The Israeli Startup Map

As of late August 2015, Israel's tech scene has the following:

- **1472** STARTUPS
- **229** SERVICES
- **84** INVESTORS
- **68** R&D CENTERS
- **61** ACCELERATORS
- **56** TECH COMMUNITIES
- **32** COWORKING SPACES

TRY EatWith

Israeli born founders Shemer Schwarz and Guy Michlin started EatWith, a company marketed as the "Airbnb for food." EatWith allows users to eat home-cooked meals around the world at people's houses—just don't make these mistakes.

Don't Bring Up the Palestine Conflict

Having just gotten to Tel Aviv, you may be thinking about all the Israel stuff you've heard about in the news, namely the Israel-Palestine conflict. But no matter whose side you pick, someone may hit you over the head with some challah in a heated debate.

Be Kosher with Kosher Foods

In America, "kosher" is an adjective usually attached to something the user deems cool, but in Israel, "kosher" refers to food following Jewish law and customs, which includes only eating animals that have cloven hooves and chew cud. Therefore, don't bring any camel or rock badger meat to someone's house. No pigs either.

Don't Overgram It

We get it. The food looks good. You want to remember it. You want to show people. We understand. But stop, put your phone in between two baklavas, and talk to people.

THE STARTUP NATION

For a country of 8.5 million people, Israel is known as "Startup Nation" for a reason. The following is the venture and exit activity during just one week in January 2015. Source: Geektime

FUNDING

- $28M Ravello Systems
- $20M AppsFlyer
- $20M EarlySense
- $18M BlueVine
- $15M TayKey
- $12M Rounds
- $8M Vaultive
- $8M RealMatch
- $6M Visualead
- $7M insert

ACQUISITIONS

- $370M Annapurna Labs acquired by Amazon
- $150M CloudOn acquired by Dropbox
- $50M Equivio acquired by Microsoft
- $200M Red Bend Software acquired by Harman

TOTALS

- $140M in funding
- $770M in acquisitions
- $910M worth of deals in one week

TIP

NETWORKING: Start holding your drink in your left hand. Having to switch hands when being introduced to someone makes you appear clumsy, and wiping your hand on your pants and offering a cold handshake is never a good first impression.

GSV'S Words to Win By

"It doesn't matter who you are, where you come from. The ability to triumph begins with you—always."

— OPRAH

"When something is important enough, you do it even if the odds are not in your favor."

— ELON MUSK

"There is nothing more powerful than an idea whose time has come"

— VICTOR HUGO

"The time is always right to do what is right."

MARTIN LUTHER KING JR.

"Imagination is more important than knowledge"

— ALBERT EINSTEIN

"The best way to predict the future is to create it."

— ABRAHAM LINCOLN

"The mind is everything. What you think you become."

— BUDDHA

"Screw it, let's do it!"

— RICHARD BRANSON

"Risk more than people think is safe. Dream more than people think is practical."

— HOWARD SCHULTZ

"Move fast and break things. Unless you are breaking stuff, you are not moving fast enough."

— MARK ZUCKERBERG

"If everything seems under control, you aren't going fast enough."

— MARIO ANDRETTI

"Why join the navy when you can be a pirate?"

— STEVE JOBS

"The greatest danger for most of us is not that our aim is too high and we miss it, but that it is too low and we reach it"

— MICHELANGELO

"Good artists copy. Great artists steal."

— PICASSO

"Where there is no vision people perish."

— PROVERBS

"Impossible is not a fact, it's an opinion."

— MOHAMMAD ALI

"Never give up. Today is hard, tomorrow will be worse, but the day after tomorrow will be sunshine."

— JACK MA

"Anything is possible if you've got enough nerve."

— JK ROWLING

"Vision without execution is hallucination."

— THOMAS EDISON

"Courage is being scared to death but saddling up anyways."

— JOHN WANYE

"You may say I'm a dreamer but I'm not the only one."

— JOHN LENNON

"Good coaches win. Great coaches cover."

— BILL CAMPBELL

11

PARIS

In Paris, extended vacation times for employees are just as common as extremely public shows of affection or words with unexpected pronunciations (e.g. rendezvous). Here, the startup atmosphere resembles the Parisian culture: slow and steady, a little pretentious, and very passionate. Etiquette and style are important, with reputations being built over lifetimes, and new models of thinking being accepted slowly. Yet perhaps no city in the world is as global as Paris. It is the center of the European Union with most European countries a train ride away—a new concept for Americans. Nowhere else in the world are there world-class cuisine, fashion, art, hustle, and inconsistent store hours in one place; the word "entrepreneur" is French for a reason.

VENTURE ACTIVITY Q4'15 - Q3'16

VC FUNDING	$1.2B
DEALS	233
EXITS	75
5 YEAR YoY FUNDING GROWTH	48.3%
5 YEAR YoY DEAL GROWTH	43.7%

Source: *CB Insights*

PARIS AT A GLANCE

Power Breakfast
- Café de la Pais
- Edgar
- Le Fouquet's
- Les Enfants Perdus
- Mini Palais

Coffee Shops to Start Up In
- Delaville
- Any Parisian café

Local Drink
- Wine

Startup District
- 2ème Arrondissement

"Must Be At" Local Events
- Bastille Day
- Circus of Tomorrow Festival
- Nuit des Musées

THE TOP

ANGELS
- Aleksandr Dresen
- Amir Banifatemi
- Fred Destin
- Scott Sage
- Xavier Niel

VCs
- Accel Partners
- Balderton Capital
- BPI France
- Daphni
- IDINVEST
- Index Ventures
- Kima Ventures
- Partech

ACCELERATORS
- 50 Partners
- Agoranov
- L'Accelerateur
- Microsoft Ventures Paris Accelerator
- Numa
- Partech Shaker
- Station F
- TheFamily

ENTREPRENEURSHIP EVENTS
- ApèrosEntrepreneurs
- HelloTomorrow
- Mash Up
- moonbar
- Start in Paris
- The Paris Founders Events

Q4'15 - Q3'16 TOP FUNDING ROUNDS
- EREN Renewable Energy—$227.8M
- Deezer—$109M
- MedDay—$38.5M
- Drivy—$35M
- Neoness—$27.3M

PARISIAN **FOOD**

The French have a lot of pride in many things, and food is the most important of them, If you can't eat like a Frenchman, your startup is doomed.

Lunch at Work Don't be that productive American who eats lunch at the cubicle, dropping a piece in between his pants and onto his chair before subtly looking around and putting it back into his mouth. If you want friends, eat with your friends.

Bonjour Say *"Bonjour"* to your waiter before you begin talking to him. His normally frowny face at the thought of catering to some tourist will light up, rainbows will appear in your wine, and the Mona Lisa will start smiling again.

Pace Speed in France is slow, so slow down. Waiters will let you set the pace, allowing them to chat and smoke a cigarette without you talking to them in broken French. Remember to ask the waiter for the *cheque* when you are finished.

Streetside cafés and restaurants are ubiquitous and perfect for anything from lunch breaks to dinner dates.

PARISIAN **CLOTHING**

If you dress like you do in Silicon Valley, Parisians will not want to be your friend. Do not wear the following:

Anything over-the-top The word "chic" originates from the French, and it is never chic to have too many sequins and psychedelic rainbow patterns on your sundress.

Anything not tight Wear tight jeans, tight hoodies, tight shirts, tight belts, but not tight fanny packs. Don't wear fanny packs. No yoga pants either. Downward Dog is not chic.

Cargo shorts It may be 95 degrees outside and you may have nothing else to wear, but underwear is better than cargo shorts. Shorts are generally a no-no as it is, but cargo shorts? The extra pockets serve no function outside of alienating you from most locals.

Flip-flops While airing out your feet, flip-flops also put a neon sign above your head that reads "Aesthetically Displeasing Slacker."

Anything that says "Paris" on it Parisians are too cool for many things, Paris being one of them.

INDEX VENTURES

French Team Members:
Martin Mignot, Rémy Luthringer

Described as sixty-year-old men with no expertise in entrepreneurship, French venture capitalists have lost out on many deals to the London offices of established venture funds like Index Ventures and Accel Partners. Here's a behind-the-scenes of Index Ventures' French investments.

Alkemics Smart-data expert for FMCG (fast-moving consumer goods) companies

Capitaine Train Train-booking tool that draws information from multiple providers

Drivy Peer-to-peer car rental service

BlaBlaCar Long-distance ride-sharing company, pairing riders who are traveling to the same far-away destination

TheFamily Respected startup accelerator that creates a community of entrepreneurs and investors

The Road from Paris

The City of Lights is only a few hours away by train or plane from the other coolest cities in Europe. Here's a quick travel guide to help you plan your jet-setting and country hopping while in town.

Xavier Niel, THE SILICON MAN

A telecommunications billionaire, Niel is a peculiar Frenchman, a self-made entrepreneur, who never studied past high school in a country where many rich Frenchmen are products of inheritance. He has complained about the French's tendency to complain and has said: "France didn't obstruct me, but it didn't help me either." With his fortune, Niel is impacting the Paris startup scene like no other Parisian before him with his incubator, school, and investments.

Station F

- Plans to open in 2018 as the "world's largest digital business incubator."
- Will host 1000 innovative startups in 33,747 square meters.
- Located in what once was the Halle Freyssinet, a transshipment hub for trucks and trains all around the country.

Kima Ventures

- Niel on startups: "It's more profitable than playing the lottery, and much more fun."
- Once boasted about investing in two startups a week.
- Cofounded with Jérémie Berrebi, an influential and serial tech investor and entrepreneur.

42

- Aims to educate 800 to 1000 students aged eighteen to thirty in programming per year.
- Named after the answer to life in Douglas Adams's *Hitchhiker's Guide to the Galaxy*.
- Students have a variety of backgrounds—40 percent have not graduated from high school while others have graduated from Stanford and MIT.
- 4.5% admission rate for the first class of students.

12

TORONTO and WATERLOO

Dubbed Silicon Pucks, Toronto and Waterloo are joined at the hip like the Valley and San Francisco, with Toronto startups having more of a web focus and Waterloo startups being more into mobile. And though startup ecosystems all around the world advertise themselves as being supportive and caring, can anyone really beat the Canadians in terms of niceness? Add to the mix a truly diverse, educated, and relaxed people who are inclusive and open for all subject matters not related to their hockey teams or superior maple syrup and poutine, and you'll understand why Y Combinator execs remarked that their most qualified YC applicants come from Waterloo. The startup scene in Toronto and Waterloo is booming—that's one thing Canadians don't have to say sorry for.

AGGREGATE VENTURE ACTIVITY Q4'15 - Q3'16

VC FUNDING	$1.7B
DEALS	225
EXITS	90
5 YEAR YoY FUNDING GROWTH	38.7%
5 YEAR YoY DEAL GROWTH	22.1%

Source: *CB Insights*

TORONTO AND WATERLOO AT A GLANCE

Power Breakfast
- Boom Breakfast & Co.
- Gabardine
- SoHo House
- Thompson Diner

Coffee Shops to Start Up In
- Bar Raval
- Early Bird Espresso
- LIT Espresso Bar
- R Squared
- Seven Grams

Best Bars
- Bar Hop
- BarVolo

Local Brew
- King Street West
- Mill Street Brew Pub
- Steam Whistle

Go-to Alcohol
- Canadian Rye

Offline Tinders
- EFS
- Lost and Found
- Maison Mercer
- The Addisons
- Wildflower

"Must Be At" Local Events
- Hockey Game
- Luminato Festival
- RedPath WaterFront Festival

THE TOP

ANGELS
- Bruce Croxon
- Jim Balsillie
- John Albright
- Perry Dellelce
- Peter Schwartz

VCs
- Extreme Venture Partners
- iNovia Capital
- OMERS Ventures
- Relay Ventures
- Round 13 Capital
- WCM Capital

ACCELERATORS
- Communitech Rev
- Creative Destruction Lab
- Highline
- INcubes
- MaRS
- Next 36
- OneEleven
- StartupNorth
- Velocity
- Utest

ENTREPRENEURSHIP EVENTS
- Camp Tech
- Enterprise Toronto
- Hack the North
- Leadership Waterloo Region
- Startup Weekend Ontario
- Techtoberfest

TOP LAWYERS
- Bennett Jones LLP
- LaBarge Weinstein
- Wildeboer Dellelce LLP

Q4'15 - Q3'16 TOP FUNDING ROUNDS
- Thalmic Labs—$120M
- Flipp—$61M
- Aeryon Labs—$45.6M
- ecobee—$35M
- Nvest—$32M

FOLLOW @THEC100 ON TWITTER

In 2010, a group of Canadian-born Silicon Valley businessmen decided to help the next generation of Canadian entrepreneurs, founding C100, which hosts a variety of networking and mentoring events for Canadian entrepreneurs. A few of their many programs:

CEO Summit Networking event between CEOs from the top growth-stage companies in Canada and influencers in Silicon Valley.

Accelerate Series Meetups for Canadian entrepreneurs and investors in many different large Canadian cities.

48hrs in the Valley Canada's most promising startups are selected to come to the Valley for 48 hours of mentorship.

Charter Member Annual Dinner Invitation-only event with a large turnout of influential Canadian entrepreneurs. C100 presents its ICE award for the leading Canadian innovator, in terms of entrepreneurship and impact on the community.

THE TOP

Hack the North, the largest hackathon in Canada, is sponsored by the engineering powerhouse University of Waterloo. Here are the some of the hacks that were created during the event:

OpenWorld Pokémon A social version of the game Pokémon, using GPS, voice control, and motion sensing to increase interactivity.

Silicon Man An iPhone mounted on a helmet that can be controlled via a glove.

RememberAll Application that works with Google Glass to get real time audio and visual feed.

Lend Airbnb for things, allowing people to rent everything from ladders to Apple TVs.

Signlang Allowing deaf people to speak aloud using Leap Motion.

Flock Facebook-connected smart door lock to control the entry and exit of people based on event attendees.

Botscape A massively multiplayer online game for artificial intelligence players. Program a bot to play the game, then bots compete against all others to highlight best programming strategies.

Spacebowl VR bowling using the Thalmic Myo and an Oculus Rift.

Guava Genomes for User Accessible Variant Annotation for secure patient genome browsing on the Web.

Pinpnt.me Web application that sends a link with your GPS location without a login or requiring personal info.

SOCIAL MEDIA CASE STUDY

A Hip Old Canadian White Man

Skyrocketing to fame with his tweets about the Meek Mill and Drake feud, Toronto councilman Norm Kelly boasts over 100,000 Twitter followers, teaching us how to effectively create a social media presence.

Norm Kelly @NormKelly
You're no longer welcome in Toronto, @MeekMill

In the Drake and Meek Kill feud, Norm chimes in, defending the honor of the Toronto-based rapper. Loyalty on social media is important.

Norm Kelly @NormKelly
I can't put out a tweet without a dozen of your replying 'dad' or 'daddy'. Why?

Twitter is a great forum to have your questions answered, showing the power of crowdsourcing on social media.

Norm Kelly @NormKelly
I won't be dropping a mixtape. Sorry.

Replying to users tagging you is very important, as good customer service is a must for any up and coming startup. Honesty is also important—here Norm makes no false promises.

Norm Kelly @NormKelly

A picture is worth a 1000 words, far more than the 140 character limit.

Norm Kelly @NormKelly
To all the young rappers who have sent me their music the past few days: Thank you but I can't get you a record deal.

Dialogue with the consumer is important. Even if they likely don't expect a response, giving consumers a response makes their day.

13

AUSTIN

Considered to be culturally rich and progressive by the majority of the United States, and full of hippies and commies by the state of Texas, Austin is a vibrant entrepreneurship city that talks and walks like San Francisco. While venture capital funding in the area pales in comparison to other startup scenes, this in turn translates to less arrogance and more helpfulness in the community, with entrepreneurship networking events every week. Add in a low cost of living, cultural events like South by Southwest, the thriving nightlife of Sixth Street, and you'll see why much of the talent that graduates from UT Austin and its CS program has no reason to leave.

VENTURE ACTIVITY Q4'15 - Q3'16

VC FUNDING	$1.5B
DEALS	286
EXITS	87
5 YEAR YoY FUNDING GROWTH	3.9%
5 YEAR YoY DEAL GROWTH	11.0%

Source: *CB Insights*

AUSTIN **AT A GLANCE**

Power Breakfast
- 24 Diner
- Counter Café
- Frank
- Anywhere with breakfast tacos

Coffee Shops to Start Up In
- 2nd Street
- Houndstooth
- Jo's Coffee

Best Pizza
- The Backspace

Best BBQ
- Franklin

Best Bars
- East Side Showroom

Local Beer
- Austin Beer Works
- Pearl Snap

Transport
- Car2Go
- Ride Austin

"Must Be At" Local Events
- ACL Music Festival
- SXSW Music/Film Festivals
- Trail of Lights

THE TOP

ANGELS
- Bradley C. Harrison
- David Cohen
- Joshua Baer
- Kenny Van Zant
- Michael Dell
- Robert Metcalfe

VCs
- Advantage Capital
- Austin Ventures
- Sevin Rosen Funds
- Silverton Partners
- Texas Emerging Tech Fund

ACCELERATORS
- Austin Technology Incubator
- Capital Factory
- Dreamit Ventures
- SKU
- TechRanch

ENTREPRENEURSHIP EVENTS
- Austin Startup Week
- Austin Tech Happy Hour
- ATX Startup Crawl
- Co-Founders Wanted
- RISE Week
- SXSW

Q4'15 - Q3'16 TOP FUNDING ROUNDS
- Civitas Learning—$60M
- Pivot3—$55M
- Spredfast—$50M
- Xerox Pharmaceuticals—$41M
- Open Lending—$40M

BALLIN' ON A **BUDGET**

Here's what to do when bootstrapping your startup...

Food Your food options are limited to Cup of Noodles (with those little plastic frozen vegetable things—you need your veggies), whatever you can scrounge from the college club meetings, and if you're Steve Jobs, the Hare Krishna Temple.

Consulting You are a startup. No matter how big your problem, please for the love of God don't pay anyone to solve it for you. As the saying goes, "A consultant is someone who borrows your watch to tell you the time and then keeps the watch."

Interns There are thousands of high school and college students who are willing to do anything for a little résumé pick-me-up, so don't hire some part-time contractor. Studies have shown that hiring an intern to fan you with palm leaves is more cost-effective than air-conditioning.

Launch parties A $5 million Series A round is great, but it does not warrant strippers, three yachts, pounds of cocaine, and flying midgets.

Share VC firms say that the surefire red flag for any prospective investment is if multiple Netflix accounts exist among a founding team. Everything needs to be shared, from access to Mom's homemade cookies to clothes to gym memberships to garage space to towels.

Coding without the Cost

Hacker Space If Steve Jobs had known about Hacker Space, an 8000-square-foot estate dedicated to any kind of builder, Apple would have been built there instead of in his garage. Hacker Space defines itself as a place for artists, computer scientists, engineers, and more who share a "hacker mind-set," defined in terms of community and a focus on intellectual growth rather than the "hacking" of security systems more often seen in the media. Hacker Space also hosts bimonthly member meetings as well as an open house every Tuesday from 8 p.m. to 11 p.m.

Capital Factory Instead of getting hellaciously wasted on Sixth Street, you can go just a little bit farther to check out Austin's most influential accelerator, located at the corner of Seventh and Brazos. While the amount of money given to accepted startups is a bit underwhelming, the community is unparalleled.

INTERACTIVE FESTIVAL TIME LINE OF THE ONE AND ONLY SXSW

A truly one-of-a-kind event, South by Southwest is split into three festivals: music, film, and interactive. While the music and film festivals are self-explanatory, the interactive festival features some of the biggest people and companies in emerging technology.

August 12, 2015
ANNOUNCING COMMUNITY GRANT FUND WINNERS Five 501(c)(3) organizations receive $10,000 grants to further their work related to one of the events in the SXSW family, from sustainability to film.

August 14, 2015
DEWEY WINBURNE AWARD NOMINEE DEADLINE The award recognizes digital do-gooders, members who help the community through new media tools and strategies.

September 4, 2015
PANELPICKER COMMUNITY VOTING CLOSES Panel Picker is an interactive online tool that allows the public to significantly impact the programming of the festival.

SURVIVING **SIXTH STREET**

Navigating Austin's hectic, rowdy nightlife.

BEFORE

POOP It's called dirty 6th for a reason, and those bathrooms go through hell and back. While many men and women have gotten away with a pee on the street, a poop is on another level.

DRINK Bars in Austin have far more reasonable prices than most big cities, but pregaming is often the most fun part of a night, with no huge crowds of people in between you and your friends.

CHARGE YOUR PHONE With the iPhone's tendency to drop from 60% to 3% when in any pocket, it is advisable to bring a portable battery as well (Anker batteries are great).

DURING

IDs While girls can get away with a pass back or a South Carolina fake, guys are in no such luck. Even if it scans and black-lights, an ID only works when accompanied by a beard.

WHAT TO DO You can dance, you can hit on boys and girls, you can drink with friends, and you can always stand, watch, and (ashamedly) realize many of these people can vote.

AFTER

FIND YOUR FRIENDS
Never has Marco Polo been more fun than when surrounded by thousands of drunk teenagers and homeless "drag rats."

RIDESHARE HOME It's never worth the risk to drive your own car to and from 6th. Don't be too drunk; there's nothing worse than spontaneously vomitting in a car that isn't yours.

October 5–7, 2015
SXSW ECO
This event encourages collaborations to solve economic, social, and environmental problems.

November 6, 2015
SXSW INTERACTIVE INNOVATION AWARDS ENTRY PROCESS CLOSES
The Awards show off the most innovative products in emerging technologies, with thirteen categories and five finalists in each category presenting at the festival.

January 16, 2016
RELEASEIT APPLICATION PROCESS ENDS ReleaseIt is a pitch competition, or "pitchathon," in the Startup Village, an entrepreneurs-only area of the festival.

March 11–15, 2016
2016 SXSW INTERACTIVE FESTIVAL

14

BANGALORE

Originally the intellectual capital of India, Bangalore, with its California feel, has become the startup capital of the subcontinent that has well over a billion people calling it home. The city has strong connections to the Valley, best exemplified in organizations like TiE and The Indus Entrepreneurs. Moreover, the amount of engineering talent in Bangalore is off the charts! Bangalore is proof that India is more than just a destination of outsourcing. Venture capital is there, and so are opportunities for networking. Huge incubators, accelerators, and workspaces are in the city, cultivating India's stars of tomorrow. What Bangalore lacks is talent in the product management and sales side of things, as well as a reliable city infrastructure—the city is known for its chaotic life when compared to the Valley, with traffic, pollution, and uncleanliness all problematic. From corruption to poverty, the problems in India are aplenty, and Bangalore hopes to provide solutions.

VENTURE ACTIVITY Q4'15 - Q3'16

VC FUNDING	$1.8B
DEALS	330
EXITS	72
5 YEAR YoY FUNDING GROWTH	65.3%
5 YEAR YoY DEAL GROWTH	47.7%

Source: *CB Insights*

BANGALORE **AT A GLANCE**

Power Breakfast
- Cafe Bateel
- Shangri La Hotel

Coffee Shops to Start Up In
- BeaglesLoft
- Costa Coffee
- Cuppa

Deal-Closing Dinners
- Karavalli
- Pink Poppadom
- Shiro

Local Beer
- Biere Club Apple Ale
- Colonial Toit
- No Parking Pilsner

Offline Tinders
- Pebble
- Skyye Lounge
- Sutra

"Must Be At" Local Events
- Bangalore Habba
- Diwali
- Ganesh Chaturthi
- Holi

THE TOP

ANGELS
- Krishnan + Meena Ganesh
- Rajan Anandan
- Sharad Sharma
- Sunil Kalra
- Vikas Taneja

VCs
- Accel Partners
- Brand Capital
- DFJ India
- Helion
- Sequoia India
- Tiger Global

ACCELERATORS
- Angel Prime
- Brand Capital/GSV
- Khosla Labs
- Kyron Accelerator
- Microsoft Ventures Bangalore
- tLabs

ENTREPRENEURSHIP EVENTS
- Eximus IIM Bangalore
- Startup Festival
- Startup Saturday
- TiE Bangalore
- VCCircle Entrepreneurs Summit

Q4'15 - Q3'16 TOP FUNDING ROUNDS
- Eurofins Scientific—$222M
- BigBasket—$150M
- Janalakshmi—$150M
- Cafe Coffee Day—$51M
- Byju's—$50M

PARENTAL **PRESSURE**

Many of Bangalore's innovators are pushed away from entrepreneurship by parents who prefer salary over equity and stability over risk. **Here's what not to tell your parents when you're about to drop out of school to pursue that social networking site for dogs.**

- There's a three percent chance this might be worth over a thousand rupees one day.
- I'd rather marry this startup than the girl you have in mind for me.
- I'm moving to Bangalore to work very hard, drink, and smoke instead of living close to you and coming to family dinners every weekend.
- I don't need a PhD to be a doctor. Look at Dr. Dre.
- Your cooking is great, but not good enough to keep me in an eight-to-six job as an accountant.

THE INDUS ENTREPRENEURS

TiE, or The Indus Entrepreneurs, is a nonprofit organization dedicated to providing mentoring, networking, and education for Indian and Indian American entrepreneurs. Here are some quick facts:

- Began in 1992 with a group of successful entrepreneurs originally from India but living in the Valley, including noted venture capitalist Kanwal Rekhi.
- Is home to 13,000 members and 2,500 charter members. There are 61 chapters in 18 countries, with a sixty-second opening in India.
- Annual TiEcon is the largest entrepreneurship conference in the world, held in Silicon Valley every May.
- TiE International Startup Competition has two tracks that allow teams to compete against others from around the world, with winners coming to the Valley for guidance and investment.
- Has youth entrepreneurship programs fostering creativity among high-school-age kids—they are planning to expand to the middle-school level as well.

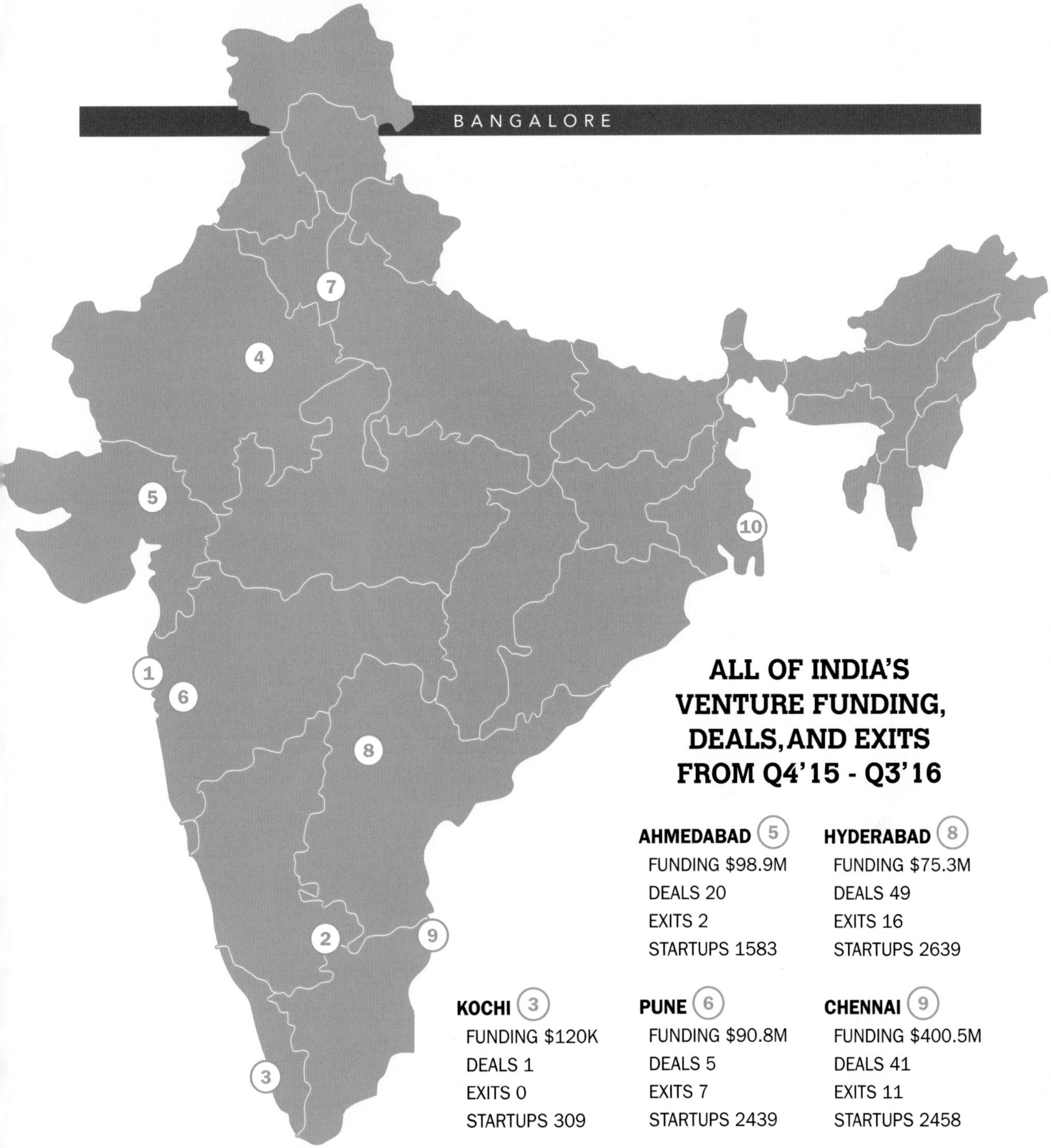

ALL OF INDIA'S VENTURE FUNDING, DEALS, AND EXITS FROM Q4'15 - Q3'16

MUMBAI (1)
FUNDING $1.2B
DEALS 233
EXITS 53
STARTUPS 5994

BANGALORE (2)
FUNDING $1.8B
DEALS 330
EXITS 72
STARTUPS 6552

KOCHI (3)
FUNDING $120K
DEALS 1
EXITS 0
STARTUPS 309

JAIPUR (4)
FUNDING $46.1M
DEALS 10
EXITS 4
STARTUPS 578

AHMEDABAD (5)
FUNDING $98.9M
DEALS 20
EXITS 2
STARTUPS 1583

PUNE (6)
FUNDING $90.8M
DEALS 5
EXITS 7
STARTUPS 2439

NEW DELHI (7)
FUNDING $3.2B
DEALS 280
EXITS 38
STARTUPS 7717

HYDERABAD (8)
FUNDING $75.3M
DEALS 49
EXITS 16
STARTUPS 2639

CHENNAI (9)
FUNDING $400.5M
DEALS 41
EXITS 11
STARTUPS 2458

KOLKATA (10)
FUNDING $11.7M
DEALS 11
EXITS 4
STARTUPS 1014

MADE IN INDIA

Some of America's biggest companies are headed by Indian men and women.

Sundar Pichai, **CEO of Google**

In 2004, Pichai joined Google at thirty-two, armed with an MS from Stanford, an MBA from Wharton, and experience at Applied Materials and McKinsey. He quickly rose through the ranks, credited with the success of Google Toolbar and Chrome, eventually becoming a part of the "L-team," executives who reported directly to Larry Page. Google employees say his meteoric rise is attributed to his empathy and kindness, which allowed him to navigate Google's politics without any enemies.

Indra Nooyi, **CEO of Pepsi**

From going to job interviews in saris because she could not afford business suits to becoming CEO of Pepsi, Nooyi remains a trailblazer for Indian women in the executive world. As Pepsi's chief strategist, she managed a series of high-profile acquisitions and deals, including the $14 billion purchase of Quaker Oats, expanding her role until taking over as CEO in 2006. For once, Pepsi is "OK."

Shantan Narayen, **CEO of Adobe**

After stints at Apple and Silicon Graphics as well as cofounding the photo-sharing website Pictra, Narayen found a home at Adobe in 1998 as a vice president. He became CEO in 2007 and pioneered Adobe's move toward the cloud for its products.

Satya Nadella, **CEO of Microsoft**

Before joining Microsoft in 1992, Satya worked at Sun Microsystems. Once at Microsoft, he held a variety of VP positions. His biggest contribution was toward Microsoft's cloud infrastructure. Nadella replaced incumbent CEO Steve Ballmer at the start of 2014, becoming the third CEO in Microsoft's history.

15

SÃO PAULO

Brazil was once the darling of the emerging market, with investors clamoring for a spot in the economy, but it has quickly fallen from good graces. While the middle class is still expanding and the economy may still boom, the optimism in the country has fallen and political corruption is a serious problem. But despite all this, the tricky labor laws, and hefty taxes, the hype began for a reason. São Paulo and Brazil have never stopped growing, the startup infrastructure is worlds ahead of where it used to be, and no one can complain about the energetic Brazilian culture, the World Cup in 2014, and the Olympics in 2016. The potential in Brazil has yet to be tapped, as championed by the story of former soccer player turned media entrepreneur Rodrigo Barros and his mission to connect Brazilian entrepreneurs to the rest of the world.

VENTURE ACTIVITY Q4'15 - Q3'16

VC FUNDING	$1.1B
DEALS	60
EXITS	34
5 YEAR YoY FUNDING GROWTH	119.0%
5 YEAR YoY DEAL GROWTH	40.4%

Source: *CB Insights*

SÃO PAULO AT A GLANCE

Power Breakfast
- Dona Deôla
- Galeria Dos Pães
- Le Pain Quotidien
- Supermercado Santa Maria
- Suplicy

Coffee Shops to Start Up In
- Coffee Lab
- Ekoa Café
- King of Fork
- Octavio Café
- Santo Grao

Startup District
- Pinheiros
- Vila Olimpia Neighborhood

Best Bar
- Frank Bar
- Octavio Café

Local Beer
- Bamberg
- Brewdog
- Invicta
- Karavelle

Go-to Drink
- Caipirinha

Offline Tinders
- Clash Club
- D.Edge
- Hot Hot
- LAB Club
- Lions
- Pub Crawl
- The Week

"Must Be At" Local Events
- Automobile Fair
- Bienal do Livro
- SPFW

THE TOP

ANGELS
- Aleksandr Dresen
- Bedy Yang
- Cassio Spina
- Dave McClure
- Eduardo Saverin
- Flavio Jansen
- Mario Ghio
- Pierre Schurmann
- Rodrigo Borges
- Romero Rodrigues
- Silvio Genesini
- Steven Vachani
- Sylvio de Barros

VCs
- Astella
- DGF
- e.Bricks
- Intel Capital
- Kaszek Ventures
- Performa
- Qualcomm Ventures
- Redpoint e.Ventures
- Ribbit Capital
- Riverwood Capital
- Rocket International
- SP Ventures
- TOTVS Ventures

ACCELERATORS
- 500 Startups
- Aceleratech
- Cubo
- Germinadora
- Startup Farm
- Wayra

ENTREPRENEURSHIP EVENTS
- BR New Tech Meetups
- Campus Party
- Dojo SP
- Silicon Drinkabout
- Startup Sampa

Q4'15 - Q3'16 TOP FUNDING ROUNDS
- Nubank—$56M
- Movile—$40M
- iFood—$30M
- Adavium Medical—$21M
- CargoX—$10M

EMERGING MARKETS VS. **SILICON VALLEY**

Outside of the obvious differences in capital and infrastructure, we take a closer look at the difference between emerging markets such as São Paulo and Silicon Valley.

ENGINEERS

The first complaint about emerging markets is usually the lack of talent. However, in these markets, engineers are not as valued as they are in the Valley, meaning attracting the best engineers is an easier task than outbidding Facebook, Google, and a host of startups. This culture of emerging markets in putting businessmen over engineers draws parallels to the culture of the Valley in the 1970s.

MOBILITY

In contrast to the American culture of moving around the country, citizens of emerging markets tend to have little geographic mobility. People tend to live and die in the same place, so in this respect, the talent isn't as concentrated.

FAMILY

Citizens in emerging markets tend to have a more family-centric life—careers, therefore, are family decisions as opposed to individual ones. Because startups have yet to be recognized as a legitimate career path, family obligations can get in the way of an enterprising entrepreneur.

VENTURE CAPITALISTS

Entrepreneurs in the Valley can expect a specialized venture capitalist for any industry, whether it be cloud, big data, social media, whatever. The emerging market VC, on the other hand, needs to be a generalist, familiar with all industries. The scene is too immature in emerging markets for specialization to be practical.

GROWTH

In Silicon Valley, nearly all high-growth companies grow with technology. In emerging markets, labor-intensive companies can have this same kind of growth curve.

CONSUMERS

Consumers in emerging markets tend to be split firmly in terms of ethnicity, region, class, and language. While developed countries are by no means homogenous, differences in these qualities are more pronounced in emerging markets.

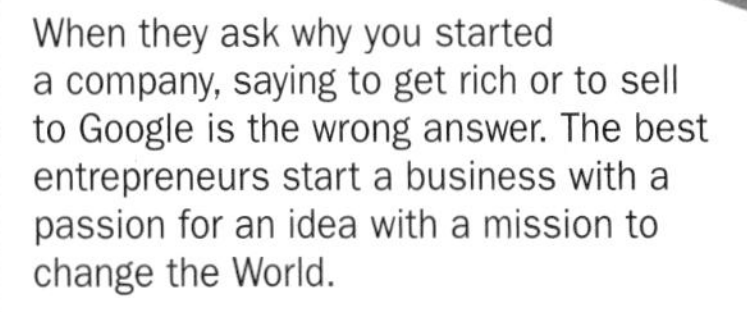

TIP

When they ask why you started a company, saying to get rich or to sell to Google is the wrong answer. The best entrepreneurs start a business with a passion for an idea with a mission to change the World.

CARNIVAL DE RIO

Held before Lent since 1723, the Carnival de Rio is one of the craziest carnivals in the world.

ACCOMMODATION

Expect airline and hotels to be way overbooked and way overpriced. If you can't book a spot with Airbnb, scouting out the best coworking places and living there for a week is always an option.

CLOTHES

Most men and women at the festival are scantily clad, especially the beautiful samba queens who lead the samba school drum sections. Drag is a perfectly normal way to go.

CULTURE

The carnival is the representation of excess, with bright colors, huge floats, and costumes everywhere. Every year the festival begins with the crowning of the Fat King, or King Momo, a tall fat man who is chosen to represent the King of Carnivals. Samba schools entertain the community with dance and music, representing a region or group of people.

Float seen during the Samba Parade at the 2007 carnival.

EDTech IN BRAZIL

Brazil's education startups are booming.

Descomplica

Online tutoring program with a focus on university admissions

Funds Raised: $13.3M
Headquarters: Rio de Janeiro

eduK

Freemium education website with a focus on entrepreneurship

Funds Raised: $10M
Headquarters: São Paulo

Geekie

Creates tools for a more personalized learning approach for students

Funds Raised: $7M
Headquarters: São Paulo

veDuca

Online video platform with videos from universities across Brazil

Funds Raised: $1.3M
Headquarters: São Paulo

Edools

Provides e-learning systems for companies

Funds Raised: $80K
Headquarters: Rio de Janeiro

Passei Direto

Connects students with study materials and other students in Brazil

Funds Raised: undisclosed, in Series B
Headquarters: Rio de Janeiro

THE WORST PLACES FOR INNOVATION

You can't have the best without also having the worst. There are some places in the world that you should avoid as a budding entrepreneur. Stay away from these innovation deadzones like the plague, unless futility and frustration are your thing.

Pyongyang, North Korea
Innovation through annihilation.

Riyadh, Saudi Arabia
Pride Rock after Scar took over.

Omaha, Nebraska
Unless you think steaks are innovative, not much happening here.

Redneck Riviera, The Gulf Coast, USA
Decades of cousins marrying cousins has depleted the IQ level.

Kentucky, USA
Ditto moonshine, which is a plenty. Venture Capital, there ain't any.

Zurich, Switzerland
Only the Swiss can make the Germans look loose.

Kabul, Afghanistan
Seventh-century "cutting edge" technology: yes; innovation: no.

Havana, Cuba
Good *cohibas*, *mojitos*, but no "*ideos*."

Montana, USA
Big sky, idea dry.

New Jersey USA
Murder, extortion, and racketeering are a buzzkill for entrepreneurs.

Jamaica
Good rum, good sun, good fun, innovation none.

The biggest hits, biggest failures, and the investors that hit the jackpot or lost the pot—these are the Unicorns and Uniscorns.

UNICORNS

Alibaba China's Amazon; had the largest IPO to date; one of the few of Yahoo!'s hits, as a VC, now worth over $230B.
  Biggest winners: SoftBank, Yahoo!

Google The Search Engine that could; now a platform that knows more about you than you know about yourself, now worth over $550B.
  Biggest winners: Sequoia, Michael Moritz, John Doerr (Kleiner Perkins)

Facebook The world's largest country with 1.7 billion "citizens," from $100 million Series A to $350+ billion market cap.
  Biggest winners: Jim Breyer (Accel), Peter Thiel

WhatsApp Acquired by Facebook for a stunning $19 billion only two years after it launched.
  Biggest winners: Jim Goetz (Sequoia)

Airbnb Allows anyone to rent out their bedroom with the hope that their guest is not a serial killer, worth over $25B.
  Biggest winner: Sequoia, Andreessen Horowitz

Dropbox Online file storage company, valued at $23 million Series A, now worth over $10 billion.
  Biggest winner: Bryan Schreier (Sequoia)

Uber The fastest-growing company in the history of technology, every cab driver's worst enemy. Valued at $60 million Series A, now worth over $60 billion.
  Biggest winner: Bill Gurley (Benchmark)

Apple When the iPod first came out, Apple's shares were $1; now Apple's worth over $600B.
  Biggest winner: Not Dell, whose namesake and chairman stated that he'd shut Apple down if he were Steve Jobs in 1997.
  NEA, Sequoia, Arthur Rock

Amazon Huge e-commerce website, driven by its powerful AWS platform. Early valuation was $60M after angel round, now the company's worth over $350B.
  Biggest winner: Kleiner Perkins, Jeff Bezos

YouTube World's largest video platform, now worth over $80 billion.
  Biggest winner: Sequoia

UNISCORNS

Webvan First Uniscorn, billion-dollar idea that went up in smoke.
Biggest loser: Sequoia, Softbank

Pets.com Typical order $30, cost to deliver $50, what's wrong with the model?
Biggest loser: Hummer Winblad Venture Partners

Kozmo Largest delivery fleet with no minimum purchase. Great for those who want a candy bar delivered to their front door.
Biggest loser: Amazon, Chase

Friendster Facebook before Zuck graduated from high school, network highjacked by Filipino housewives.
Biggest loser: Benchmark, Kleiner Perkins, Battery

Solyndra Permanently and forever taking Uncle Sam out of the VC business
Biggest loser: U.S. Department of Energy

Pay By Touch Biometrics = bye bye money
Biggest loser: Mobius Venture Capital

Boo.com British Internet company found by Swedes, who earned their domain name.
Biggest loser: Goldman Sachs, Amazon, Bernard Arnault

eToys Great idea if toddlers were on the Internet and had credit cards.
Biggest loser: Intel, Sequoia

Better Place $800 million vanished, bankruptcy put investors out of misery and in a better place.
Biggest loser: GE, Morgan Stanley, VantagePoint

CueCat Backed by RadioShack, two wrongs don't make a right.
Biggest loser: RadioShack, but begs the question, if you're already lost, can you lose?

16

SEATTLE

Seattle could very well be construed as the hip but perpetually rainy headquarters of the supercorporations Microsoft, Starbucks, and Amazon, but there is more to Seattle than that. Many in Seattle would rather take the comfortable six-figure salary over the path of a founder. Those who decide not to will find a more laid-back community that boasts multi-billion-dollar exits in the likes of Tableau and Zulily. This culture of contentment rather than competition has translated into more customer-oriented companies, such as Costco and Starbucks, with Nordstrom as the pioneer of the consumer culture. Seattle is not the Valley, nor does it want to be.

VENTURE ACTIVITY Q4'15 - Q3'16

VC FUNDING	$866.7M
DEALS	277
EXITS	60
5 YEAR YoY FUNDING GROWTH	11.9%
5 YEAR YoY DEAL GROWTH	7.1%

Source: *CB Insights*

SEATTLE AT A GLANCE

Power Breakfast
- Hi-Spot Cafe
- Portage Bay Café
- Wandering Goose

Coffee Shops to Start Up In
- Analog
- Broadcast
- Caffe Ladro
- Herkimer
- Starbucks Madison Park
- Tully's Clyde Hill

Best Bar
- Brave Horse Tavern
- Shelter

Local Beer
- Dubbel Entendre
- Mack & Jack
- Rainier

"Must Be At" Events
- Bumbershoot
- Fremont Solstice
- Seattle Seafair
- Washington State Fair

In the Know

Howard Schultz is not only the mayor of Seattle, but also the Governor of Washington and potentially the POTUS in 2020.

THE TOP

ANGELS
- Andy Sack
- Ben Slivka
- Chris DeVore
- Gary Rubens
- Geoff Entress
- Hadi Partovi
- Rudy Gadre

VCs
- Divergent
- Frazier Healthcare
- Ignition Partners
- Madrona
- Maveron
- Second Avenue
- Triilogy
- Vulcan
- WRF Capital

ACCELERATORS
- CoMotion
- Fledge
- Reactor
- Techstars

ENTREPRENEURSHIP EVENTS
- IT Summit Seattle
- Seattle Angel Conference
- Seattle Startup Weekend
- Seattle Tech Leadership Summit
- UW Business Plan Competition

Q4'15 - Q3'16 TOP FUNDING ROUNDS
- DocuSign—$300M
- Wave Broadband—$125M
- OfferUp—$119M
- Avalara—$96M
- Kymeta—$62M

17

SINGAPORE

There is more access to capital in this city of many rules than ever before—as best represented with Softbank's $250 million investment in GrabTaxi. In the Asia startup scene, India and China are the clear leaders, but despite its small market, Singapore's infrastructure is world-class and closely resembles Israel—its "rags to riches" story is one of imposing its will to achieve success. Block 71 and Block 79 are considered the core of the entrepreneurship scene. Look up *get shit done* in our glossary and you'll see a picture of Singapore.

VENTURE ACTIVITY Q4'15 - Q3'16

VC FUNDING	$2.4B
DEALS	180
EXITS	67
5 YEAR YoY FUNDING GROWTH	38.0%
5 YEAR YoY DEAL GROWTH	22.2%

Source: *CB Insights*

SINGAPORE **AT A GLANCE**

Power Breakfast
- Café Gavroche
- Edge

Coffee Shops to Start Up In
- Kith Café
- Patisserie G

Local Beer
- Pump Room IPA

"Must Be At" Local Events
- Chingay Parade
- Marina Bay on New Year's
- Zoukout

In the Know

Local development VC fund EDBI is the "fast track" for entrepreneurs who can benefit Singapore.

THE TOP

ANGELS
- Dusan Sankovic
- Eduardo Saverin
- Jayesh Parekh
- Khailee Ng
- Peng T Ong

VCs
- EDBI
- GIC
- Golden Gate Ventures
- Rakuten
- Sequoia Capital
- Temasek
- Vertex Ventures
- Wavemaker Partners

ACCELERATORS
- Fatfish Media Lab
- iAxil
- ImpactHub
- Joyful Frog
- Singapore Infocomm Technology Federation
- Working Capitol

ENTREPRENEURSHIP EVENTS
- Singapore Startup Weekend
- Startup Asia
- Tech in Asia
- TiE
- Youth Entrepreneurship Symposium
- YouthHack

Q4'15 - Q3'16 TOP FUNDING ROUNDS
- Lazada—$1B
- Olam International—$650M
- QuEST Global Services—$350M
- e-Shang Redwood—$300M
- Trax Image Recognition—$40M

Singapore's Startup Ecosystem

Unlike other cities in the world, the government of Singapore is the main ally of the startups in the city.

Spring Singapore

A*STAR

EMI

18

AMSTERDAM

There is more to Amsterdam than the red-light district and legal marijuana. It boasts an excellent education system and a government interested in tech entrepreneurship. The Highly Skilled Migrant Visa and 30% Ruling helps skilled immigrants come into the city, and Amsterdam's affordability, open culture, and English-speaking population means that they want to stay. Access to capital and seasoned mentors is still a problem, but entrepreneurship has been and remains a tradition in Amsterdam.

VENTURE ACTIVITY Q4'15 - Q3'16

VC FUNDING	$189.4M
DEALS	111
EXITS	29
5 YEAR YoY FUNDING GROWTH	13.9%
5 YEAR YoY DEAL GROWTH	37.0%

Source: *CB Insights*

AMSTERDAM AT A GLANCE

Power Breakfast
- Lloyd Hotel
- Café-Restaurant de Plantage
- The Lobby

Coffee Shops to Start Up In
- Coffee Co.
- La Place
- The Coffee Virus

Local Beer
- Verdoemenis

Offline Tinder
- CafeLux

Best Bar
- Vesper

"Must Be At" Local Events
- Grachtenfestival (Canal Festival)
- International Domestic Film Festival
- King's Birthday

THE TOP

ANGELS
- Boris Veldhuijzen van Zanten
- Jonathan Nelson
- Raj Ramanandi
- Rune Theill

VCs
- HENQ Invest
- NPM Capital
- Peak Capital
- Prime Ventures
- SET Ventures
- StartGreen Capital
- Van den Ende & Deitmers

ACCELERATORS
- ACE Venture Lab
- Founder Institute
- RockStart
- StartupBootCamp

ENTREPRENEURSHIP EVENTS
- ICEID Amsterdam
- Startup Weekend Amsterdam
- Techallstars
- The Next Web Conference
- TNW Amsterdam

Q4'15 - Q3'16 TOP FUNDING ROUNDS
- Pyramid Analytics—$30M
- Avantium Technologies—$23M
- Binder Image Library—$22M
- 3D Hubs—$7M
- BUX—$6.9M

19

DENVER and BOULDER

The Denver and Boulder ecosystem is still looking for its juggernaut, but young burnouts from New York and Silicon Valley have found an actual work-life balance in the area. With a young and progressive population, affordable living, sunny weather, and a variety of natural beauty, there is a much more laid-back startup culture in Colorado. This leads to a more helpful and collaborative atmosphere—startups are sharing networks, as opposed to poaching employees. 10,000 people a month are moving to Denver to take advantage of the sun, the healthy lifestyle, the lower cost of living and the startups.

AGGREGATE VENTURE ACTIVITY Q4'15 - Q3'16

VC FUNDING	$2.5B
DEALS	285
EXITS	83
5 YEAR YoY FUNDING GROWTH	14.0%
5 YEAR YoY DEAL GROWTH	11.1%

Source: *CB Insights*

DENVER AND BOULDER AT A GLANCE

Power Breakfast
- Denver Biscuit Company
- Lucile's
- Perfect Landing
- Snooze

Coffee Shops to Start Up In
- Amante Coffee
- Huckleberry Roasters
- Ozo Coffee Pearl
- The Cup

Startup District
- LoHi
- Pearl Street

Best Bars
- Forest Room 5
- Highland Tavern
- Pearl Street Pub
- The Bitter Bar
- The West End Tavern

Local Brewery
- Avery Brewing Co.
- Ratio Beerworks
- Upslope Brewing Co.
- Wynkoop Brewing Co.

"Must Be At" Events
- Colorado 420 Weekend
- Riot Fest and Rodeo
- The Boulder Creek Festival

In the Know

The hottest social advertising firm on planet, Made Movement, is based in Boulder.

THE TOP

ANGELS
- Bart Lorang
- Brad Feld
- James Franklin
- Jason Seats
- Nicole Glaros

VCs
- Boulder Ventures
- Drummond Road Capital
- Foundry Group
- High Country Venture
- Sequel Venture Partners

ACCELERATORS
- BoomTown
- Galvanize
- Innovation Center of the Rockies
- Innovation Pavilion
- MergeLane
- Techstars
- Unreasonable Institute

ENTREPRENEURSHIP EVENTS
- Aspen Ideas Festival
- Boulder Beta
- DeFrag Conference
- Fortune Tech Brainstorm
- New Tech Colorado
- PitchSlam
- Startup Week

Q4'15 - Q3'16 TOP FUNDING ROUNDS
- SomaLogic—$60.5M
- PanTheryx—$53M
- LogRhythm—$50M
- Datavail—$47M
- Galvanize—$45M

20

BALTIMORE–DC–VA METRO AREA

Sure, it's not the most specific destination, but the entrepreneurship scenes of Baltimore, DC, and Northern Virginia are closely connected. They benefit from proximity to the federal government, as these startups can contribute to national security and governmental technology without the bureaucracy. The area's network of colleges allows for a constant stream of talented young people, and the region most resembles a small-scale New York with more cautious investors and a high concentration of education and health startups.

AGGREGATE VENTURE ACTIVITY Q4'15 - Q3'16

VC FUNDING	$1.7B
DEALS	300
EXITS	69
5 YEAR YoY FUNDING GROWTH	10.7%
5 YEAR YoY DEAL GROWTH	11.3%

Source: *CB Insights*

BALTIMORE–DC–VA METRO AREA **AT A GLANCE**

Power Breakfast
- Farmers Fishers Bakers
- Four Seasons
- Le Diplomate

Coffee Shops to Start up In
- Bourbon
- Compass
- Philz
- Potter's House

Best Bar
- ANXO Cidery & Pintxos Bar
- Barcelona on 14th
- Dacha in Shaw
- Jack Rose
- 1Right Proper
- The Passenger
- The Soverign

Local Beer
- DC Brau
- Natty Boh

"Must Be At" Local Events
- Atlantic Ideas Fest
- Independence Day Festivities
- National Cherry Blossom Festival
- Nationals Games
- Passport DC

THE TOP

ANGELS
- Bobby Yazdani
- Jay Virdy
- Michael Chasen
- Miles Gilburne
- Nigel Moms
- Sean Glass
- Steve Case
- Ted Leonsis
- Tony Florence

VCs
- ABS Capital
- Grotech
- In-Q-Tel
- JMI
- NEA
- Novak Biddle
- Revolution
- T. Rowe Price

ACCELERATORS
- 1776
- Acceleprise
- bwtech@UMBC
- Mach 37

ENTREPRENEURSHIP EVENTS
- ACS Entrepreneurial Summit
- America's Small Business Summit
- Startup Challenge Cup
- Women Leading the Future Summit

Q4'15 - Q3'16 TOP FUNDING ROUNDS
- Precision for Medicine—$75M
- RainKing—$67M
- Higher Logic—$55M
- Lookingglass—$50M
- GrayBug—$45M

DC VC— 1776

With roots in Washington, DC, 1776 is a global incubator and seed fund. Its annual event, the Challenge Cup, is a worldwide tournament for the most promising startups to win prizes, network, and share their ambitious vision with the rest of the world. The Challenge Cup consists of 45 local competitions, 9 regional competitions, and 1 Global Final.

21

MIAMI

Of all the American startup ecosystems, Miami is perhaps the most international, as it is heavily intermixed with the Latin American tech landscape. Miami boasts a variety of industries and a bustling nightlife, and people are constantly taking their talents to South Beach, even though technical talent remains a pressing need. Neighboring cities Boca Raton, Fort Lauderdale, Dania Beach, and Hollywood also draw in significant VC interest. Though capital continues to increase, ample funding past the seed round is still hard to come by. The northern capital of South America is also the birthplace of the Cuban Sandwich, and it has only snowed once in Miami's meteorological history.

AGGREGATE VENTURE ACTIVITY Q4'15 - Q3'16

VC FUNDING	$756.9M
DEALS	136
EXITS	84
5 YEAR YoY FUNDING GROWTH	44.9%
5 YEAR YoY DEAL GROWTH	17.0%

Source: *CB Insights*

MIAMI **AT A GLANCE**

Power Breakfast

- Big Fish
- Icebox Café
- Miami Café
- Yardbird

Coffee Shops to Start Up In

- Angelina's
- Panther Coffee

Local Beer

- Miami Weiss

Startup Districts

- Brickell
- Midtown
- Wynwood

Best Bars

- Batch Gastropub

"Must Be At" Local Events

- Art Basel
- Art Deco Weekend Festival
- SOBE Food and Wine

In the Know

Don't wear a Castro shirt in Little Havana.

THE TOP

ANGELS

- Gonzalo Costa
- Hector Hulian
- Marcelo Claure
- Mark Kingdon
- Peter Kellner
- Shaun Abrahamson

VCs

- Athenian Venture Partners
- H.I.G. Capital
- Knight Enterprise Fund
- Ladenburg Thalmann
- Medina Capital

ACCELERATORS

- Endeavor
- Idea Center at Miami Dade
- Lab Miami
- Rokk3r Labs
- Venture Hive
- Winced

ENTREPRENEURSHIP EVENTS

- Emerge Miami
- OpenHack Miami
- Startup Grind Miami
- Startup Weekend
- Tech Night at The BallPark

Q4'15 - Q3'16 TOP FUNDING ROUNDS

- Magic Leap—$827M
- Finova Financial—$52.5M
- Hygea Holdings—$40M
- InnFocus—$33.9M
- Airspan—$30M
- JetSmarter—$30M

22

MUMBAI

The city formerly known as Bombay is similar to Bangalore in some aspects. The cities share a high concentration of technical talent thanks to the India Institutes of Technology (IIT) system, but both have city infrastructures that leaves much to be desired. In contrast to Bangalore, however, Mumbai startups have a greater focus on products as opposed to IT services. The obsessions with Bollywood and mobile phones fuel opportunities in entertainment, games, and social media. Mumbai is also the finance and media capital of India, meaning there is a much more diverse talent pool.

VENTURE ACTIVITY Q4'15 - Q3'16

VC FUNDING	$1.2B
DEALS	233
EXITS	53
5 YEAR YoY FUNDING GROWTH	10.7%
5 YEAR YoY DEAL GROWTH	42.3%

Source: *CB Insights*

MUMBAI **AT A GLANCE**

Power Breakfast
- Bagel Shop
- Salt Water
- Sassanian Boulangerie

Coffee Shops to Start Up In
- Barista
- Gaylord
- Mocha

Local Beer
- Barking Deer
- Bombay Blonde
- Flying Pig

"Must Be At" Local Events
- Cheer!
- Kala Ghoda Arts Festival
- Prithvi Theatre Festival

In the Know

The iconic Taj Hotel is located across the "Gateway of India" and continues to serve as a place to do business for locals and visitors alike.

THE TOP

ANGELS
- Anand Ladsariya
- Anirudh Damani
- Mahesh Murthy
- Ravi Kiran
- Sanjay Kamlani

VCs
- Bessemer
- Brand Capital
- Mayfield
- Nirvana Venture Advisors
- Norwest Venture Partners

ACCELERATORS
- GSF
- Seedfarm
- SINE
- UnLtd
- VentureNursery

ENTREPRENEURSHIP EVENTS
- Ignite
- Startup Weekend Mumbai
- Techfest
- Techsparks
- The Entrepreneurship Summit

Q4'15 - Q3'16 TOP FUNDING ROUNDS
- Ola—$500M
- India Infoline—$173M
- CarTrade.com—$145M
- Crompton Greaves—$126M
- BookMyShow—$81.5M

23

ATLANTA

Southern hospitality is everywhere in the commerce hub of the South. There truly is a tight-knit startup scene anchored by Georgia Tech. And since it's a smaller pond than the ocean that is the Valley, you don't need to have founded Facebook to attract the best talent—and you don't need a couple million dollars to buy a house. Though the scene is still scattered and the low cost of living means startups doomed to fail crawl to their grave far too slowly, Atlanta does have talented startups in ad-tech, security, and health, as well as big names in MailChimp, whose mispronounced ad before every *Serial* podcast you probably remember.

VENTURE ACTIVITY Q4'15 - Q3'16

VC FUNDING	$1.7B
DEALS	133
EXITS	64
5 YEAR YoY FUNDING GROWTH	26.6%
5 YEAR YoY DEAL GROWTH	2.3%

Source: *CB Insights*

ATLANTA **AT A GLANCE**

Power Breakfast
- Babs
- Corner Café
- Folk Art
- Gato
- Juliana's
- Highland Bakery
- Sublime Donuts
- Sun in My Belly
- Waffle House

Coffee Shops to Start Up In
- Atlanta Coffee Roasters
- Octane

Best Bar
- 5 Seasons
- Blind Willie's
- Cypress Pint and Plate

Local Beer
- 420 Extra Pale Ale
- Blind Willie's
- Monday Night Brewing
- Wild Heaven

"Must Be At" Local Events
- Chamblee Antique Row
- Dogwood Festival
- Dragon Con
- Inman Park Festival
- Music Midtown
- Woodruff Park

In the Know

Atlanta is home to the world's busiest airport, and the city is a convenient three hours from most major American cities.

THE TOP

ANGELS
- David Cummings
- Jason Seats
- Jimmy Liu
- Jon Gordon
- Paul Judge
- Sig Mosley

VCs
- BIP Capital
- Cox Venture Fund
- Forte Ventures
- Fulcrum Equity Partners
- Kinetic Ventures
- Noro-Moseley Partners
- TechOperators
- Tech Square Ventures

ACCELERATORS
- ATDC—Advanced Technology and Development Center
- Atlanta Ventures Accelerator (Atlanta Tech Village)
- Flashpoint
- Switchyards

ENTREPRENEURSHIP EVENTS
- Digital Summit
- Govathon
- HackGT
- Startup Chowdown
- Startup Lounge
- Venture Atlanta

Q4'15 - Q3'16 TOP FUNDING ROUNDS
- Carvana—$160M
- Kabbage—$135M
- Ankura Consulting Group—$100M
- Pindrop Security—$75M
- RiverMend Health—$60M

24

SEOUL

Barring a Kim Jong-un tantrum-produced missile attack, the South Korean startup scene's potential is off the charts. The Gangnam District in Seoul is the center of innovation, and its "Gangnam Style" is just taking off. Seoul is still unproven, with bankruptcy laws that do not help founders, but the South Korean infrastructure is beyond developed—80% of the country has smartphones and Seoul's WiFi is blisteringly fast. Add in the South Korean government's $3.7 billion dollar investment in startups, and you'll see why 500 Startups has an office in Seoul called 500 Kimchi.

VENTURE ACTIVITY Q4'15 - Q3'16

VC FUNDING	$456.4M
DEALS	39
EXITS	13
5 YEAR YoY FUNDING GROWTH	90.3%
5 YEAR YoY DEAL GROWTH	67.0%

Source: *CB Insights*

SEOUL AT A GLANCE

Power Breakfast

- Bistro Miru
- Gusto Taco
- Hemlagat

Coffee Shops to Start Up In

- Charlie Brown Café
- Coffeest
- Piano Café

Local Beer

- Weizen at Queenshead

"Must Be At" Local Events

- Korea International Art Fair
- Lotus Lantern Festival
- Morronier Summer Festival

In the Know

Don't decline the first round of drinks that are poured for you.

THE TOP

ANGELS

- Benjamin Joffe
- Christine Tsai
- Jimmy Rim
- Net Jacobsson
- Tim Chae

VCs

- Insight
- K Cube Ventures
- KSLSF
- Softbank Ventures
- Stonebridge Capital

ACCELERATORS

- Fast Track Asia
- Future Play
- K-Startup
- Sparklabs
- Venture Square

ENTREPRENEURSHIP EVENTS

- beGlobal Seoul
- beLaunch
- Business Network Korea
- Max Summit
- Seedstars Seoul
- Startup Grind Seoul

Q4'15 - Q3'16 TOP FUNDING ROUNDS

- Webzen—$170M
- YG Entertainment—$85M
- Memebox—$65.9M
- Baedal Minjok—$50M
- Wooza Bros.—$50M

25

STOCKHOLM

Once known almost exclusively as the birthplace of ABBA, Stockholm and Sweden are now responsible for Skype, Spotify, and the two most addicting games around: Candy Crush Saga and Minecraft. The blue-eyed, blond-haired, and at times blue-eared (it's cold out there) people have some very talented programmers, a cashless society, and a virtue of humility called *"Jantelagen"*: a stark contrast to the name-dropping, constantly advertising people of Silicon Valley. Startups in Stockholm grow quickly because of the city's early adopters and global perspective.

AGGREGATE VENTURE ACTIVITY Q4'15 - Q3'16

VC FUNDING	$1.4B
DEALS	143
EXITS	50
5 YEAR YoY FUNDING GROWTH	44.1%
5 YEAR YoY DEAL GROWTH	53.1%

Source: *CB Insights*

STOCKHOLM **AT A GLANCE**

Power Breakfast
- Grand Hotel
- Riche
- Zink Grill

Coffee Shops to Start Up In
- Coffice
- Drop Coffee
- Vete-Katten

Local Beer
- Carnegie Porter
- Crocodile
- Falcon
- Påsköl

Offline Tinder
- Trädgården

"Must Be At" Local Events
- Liquorice Festival
- Malmö Festival
- Stockholm Pride

In the Know

In Sweden, happy hour is called "after work." Perhaps that's their secret for their long life expectancy.

THE TOP

ANGELS
- Anil Hansjee
- Floris Rost van Tonningen
- Net Jacobsson
- Roberto Bonanzinga
- Sean-Seton Rogers

VCs
- Creandum
- Edastra
- HealthCap
- Industrifonden
- Investor AB
- Northzone

ACCELERATORS
- Bonnier Accelerator
- Epicenter
- Frogleap
- Shift
- STING
- Stugan
- SUP46

ENTREPRENEURSHIP EVENTS
- Bootcamp
- Meetup Stockholm
- Startup Day
- Startup Weekend
- STHLM TECH FEST

Q4'15 - Q3'16 TOP FUNDING ROUNDS
- Spotify—$1B
- Pomegranate—$67.6M
- Starbreeze—$40M
- iZettle—$64M
- Klarna—$30M

BEST AND WORST ACQUISITIONS

When Facebook acquired Instagram for $1 billion, everyone questioned the purchase...but it turned out to be a huge hit for both companies, as Instagram is now valued at over $35 billion. Unfortunately, that doesn't hold true for Yahoo!'s multibillion-dollar acquisition of Mark Cuban's now defunct Internet radio company. Here are a few of the biggest acquisition hits and misses made by technology companies in the past few years.

WORST

Broadcast.com bought by Yahoo! in 1999 for $5.7 billion.
Yahoo!'s ROI was 100%...that is, if ROI = Radio on the Internet.

Geocities bought by Yahoo! in 1999 for $3.6 billion.
Shut down in 2009, further cementing Yahoo!'s highly volatile history as a VC.

MySpace bought by NewsCorp in 2005 for $520 million.
The once white-hot social network was sold for $35 million in 2011.

Hotmail bought by Microsoft in 2005 for $400 million.
An e-mail platform that didn't live up to its name.

Palm bought by HP for $1.2 billion in 2010.
HP put Palm out of its misery and shut down production of all devices in 2011.

Motorola bought by Google in 2012 for $12.5 billion.
A rare stumble for Google, Motorola was sold two years later writing down their investment by 75%.

OMGPOP bought by Zynga for $180 million in 2012.
Zynga pulled an OMG and POP'd its acquisition, shutting it down in 2013.

Lycos bought by Terra Networks for $12.5 billion in 2000.
Sold for $105 million in 2004...less than 2% of the multibillion-dollar investment.

BEST

Instagram bought by Facebook in 2012 for $1 billion when it had 15 employees.

Now estimated to be worth $35+ billion, Instagram is *the* go-to photosharing platform for celebs, teens, and creepers.

Android bought by Google in 2005 for $50 million.

Now installed on over 1.4 billion mobile devices and in almost all countries, it is the dominant mobile OS.

YouTube bought by Google in 2006 for $1.65 billion.

Now the biggest video platform in the world, estimated to be worth $80 billion.

DoubleClick bought by Google for $3.1 billion in 2007.

Platform that drives Google's online advertising, and thus Google's economic engine.

DOS bought by Microsoft for $75,000 in 1981.

DOS eventually became MS DOS...which led to a little operating system called Microsoft Windows.

Lenovo bought by IBM in 2005 for $1.25 billion.

Drove Lenovo's business to new heights and a $10 billion market cap.

PayPal bought by eBay in 2002 for $1.5 billion

Online payment platform meet online marketplace; a match made in heaven that drove PayPal to a $35 billion market cap.

Zappos bought by Amazon in 2009 for $1.2 billion.
Zappos brought both sole and soul to Amazon.

VMware bought by EMC in 2004 for $625 million.

The company is now worth $20.6 billion.

Steve Jobs (aka NeXt Computer), bought by Apple in 1996 for $429 million.
See Glossary entry "Steve Jobs."

In 1996, Apple's market cap was $3.1 billion. Today, it is now over $500 billion.

TIP

While it's true that Steve Jobs liked to go barefoot, wearing shoes and having good hygiene is a better strategy for capital-seeking entrepreneurs. You can become eccentric once you're rich, but this is an area where you don't fake it before you make it.

The Pioneer Series

Marc Benioff

- Founded Salesforce.com in 1999 with the sole mission to revitalize the software industry to revamp the way that software programs were designed and distributed.
- At Salesforce, Benioff pioneered the 1/1/1 philanthropic model by which companies contribute 1% of profits, 1% of equity and 1% of employee hours back to the community it serves.

Bill Campbell—Coach of the Valley

- Key advisor to the founders and CEOs of Amazon, Google, Twitter, and countless Silicon Valley companies.
- Former Apple board member and CEO/chairman of Intuit.

Vinod Khosla

- Cofounder of Sun Microsystems and served as the first chairman and CEO of the company.
- Former general partner at Kleiner Perkins, and founded Khosla Ventures in 2004, which has since become one of the most recognized venture capital firms in the world.

Benchmark Boys—Bill Gurley, Mitch Lasky, Matt Cohler, Peter Fenton, Eric Vishria, Scott Belsky

- One of the top venture capital firms in the Valley, with early-stage investments in numerous successful startups such as Dropbox, Uber, Twitter, Snapchat, and Instagram, and the six of them serve as board members for such companies.

Vinod Khosla speaking at the 2015 ASU GSV Summit

Diane Greene

- Cofounder, president and CEO of VMware, one of the largest virtualization companies in the world, which was acquired by EMC.
- Currently the senior vice president for Google's cloud businesses and serves on the boards of MIT and Khan Academy.

Larry Sonsini

- Chairman of Wilson Sonsini, one of the preeminent law firms in the Valley.
- Advised companies such as Google, Apple, Sun Microsystems, Netscape, and YouTube.

Mark Zuckerberg

- Founded Facebook, which is now the largest social communication platform in the world with over 1.6 billion users, in his dorm room.
- Pledged to give 99% of his Facebook shares, worth about $45 billion, toward the Chan Zuckerberg Initiative.

RON CONWAY

Arguably the most prolific angel investor in Silicon Valley, RON CONWAY is legendary for his ability to connect to people and make friends. His investments include early financings of Google, Facebook, Pinterest, and Twitter, and his friends include everyone from Marissa Mayer to Brian Wilson to MC Hammer. Since the mid-1990s, Conway has invested in over 600 Internet companies, and is considered to be the pioneer of angel investing. There is the right way, the wrong way, and the Conway.

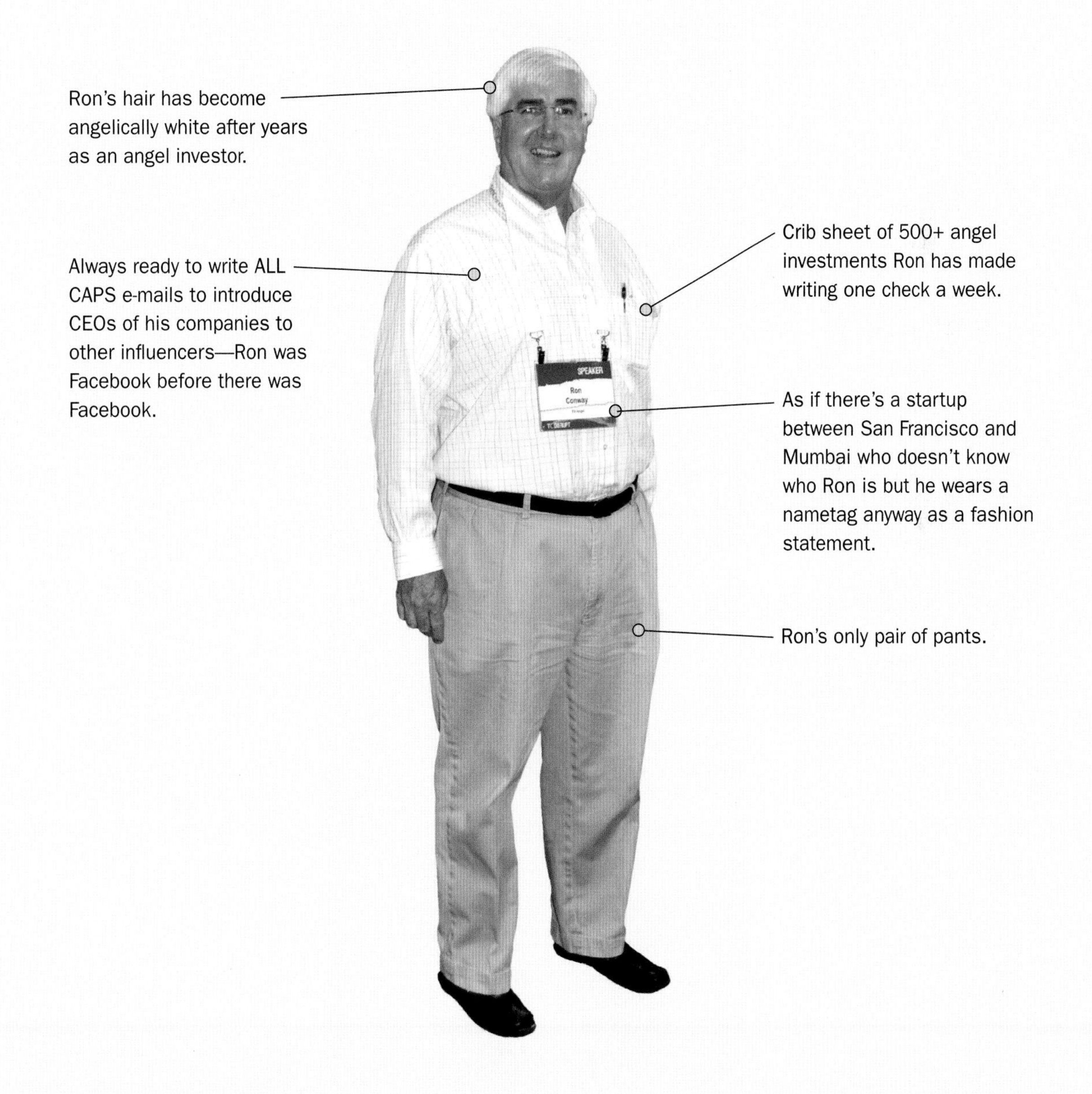

26

NEW DELHI and GURGAON

The capital of India has some of the largest startups in the land of Gandhi. Neighboring city Gurgaon has also become a center of tech activity, and is home to over 250 of the Fortune 500. Rock star prime minister Narendra Modi launched "Make in India" in 2014 to encourage multinational companies to manufacture products in India.

AGGREGATE VENTURE ACTIVITY Q4'15 - Q3'16

VC FUNDING	$3.2B
DEALS	280
EXITS	38
5 YEAR YoY FUNDING GROWTH	83.9%
5 YEAR YoY DEAL GROWTH	68.5%

Source: *CB Insights*

NEW DELHI AND GURGAON **AT A GLANCE**

Power Breakfast
Sagar Ratna

Coffee Shops to Start Up In
Café Coffee Day
Glenz Café n Bakers

Local Beer
Kingfisher

"Must Be At" Local Events
Shankar Market Connaught Place New Delhi
Going to the blueFROG
International Mango Festival

In the Know
The entire public transport system runs on eco-friendly compressed natural gas, making the city one of the greenest in the world.

THE TOP

ANGELS
Aloke Bajpai
Anirudh Mullick
Jay Meattle
Rakesh Agrawal
Sandeep Murthy

VCs
Brand Capital
Canaan Partners
Indian Angel Network
Lightspeed Ventures
Sequoia Capital India
SIDBI Venture Capital

ACCELERATORS
Brand Capital/GSV
Entreprensurship Development Cell IIT Delhi
Indian Angel Network
TLabs

ENTREPRENEURSHIP EVENTS
Startup Weekend New Dehli
Techcircle Startup
The Startup Garage
TiE Delhi

Q4'15 - Q3'16 TOP FUNDING ROUNDS
ReNew Power—$265M
ibiboGroup—$250M
MakeMyTrip—$180M
Hike—$175
Welspun Renewables—$165M

DALLAS and FORT WORTH

The Big D is headquarters for dozens of Fortune 500 companies and billionaires, including Global Silicon Valley celebrity Mark Cuban. The startup scene here is fragmented, mainly due to the distance between the two large cities, but hey, everything's bigger in Texas, ya'll. Given the density of major corporations and no state income tax, Texans with a wildcatter attitude do the hoedown in the Metroplex.

AGGREGATE VENTURE ACTIVITY Q4'15 - Q3'16

VC FUNDING	$2.2B
DEALS	129
EXITS	82
5 YEAR YoY FUNDING GROWTH	5.2%
5 YEAR YoY DEAL GROWTH	13.3%

Source: *CB Insights*

DALLAS AND FORTH WORTH **AT A GLANCE**

Power Breakfast
- Bolsa
- Crossroads Diner
- Jonathon's Oak Cliff

Coffee Shops to Start Up In
- Local Press + Brew
- Mudsmith
- Oak Lawn Coffee
- Opening Bell Coffee

Best Bar
- Joule
- Midnight Rambler

"Must Be At" Local Events
- DIFF
- St. Patrick's Day Parade and Festival
- State Fair of Texas
- TEDxSMU

In the Know

Over twenty-five billionaires call Dallas home, with the tribe led by Mark Cuban.

THE TOP

ANGELS
- Alex Chang
- Kevin Moore
- Mark Cuban
- Morton Meyerson
- Raju Indukuri

VCs
- Dallas Venture Partners
- Green Park & Golf Ventures
- Hangar Ventures
- Naya Ventures
- Outlier Ventures

ACCELERATORS
- Health Wildcatters
- REVTECH
- Tech Wildcatters
- The GroundFloor @ United Way

ENTREPRENEURSHIP EVENTS
- Digital Dallas
- Startup Career Fair
- Startup Grind
- Startup Week

Q4'15 - Q3'16 TOP FUNDING ROUNDS
- RimRock Oil and Gas—$500M
- Lucid Energy Group—$350M
- StackPath—$180M
- Teal Natural Resources—$125M
- Peloton Technologies—$52.4M

28

DUBLIN

Dublin, nicknamed Tech City, is the largest city in Ireland and one of the liveliest cities in the world. That reputation has been emboldened by the "happening" Web Summit, born in Dublin but has since outgrown itself to live in Lisbon, and contains a rich history of fun—and it is held at the first pub in history. With the youngest population in all of Europe, Dublin is home to 666 licensed bars. Half of the population is younger than twenty-five, which makes the drinking age of eighteen convenient. As anecdotal evidence for alcohol's enlightenment effect, Dublin's famous Trinity College boasts many celebrated graduates, including Oscar Wilde, Jonathan Swift, and Bram Stoker.

VENTURE ACTIVITY Q4'15 - Q3'16

VC FUNDING	$1.6B
DEALS	108
EXITS	52
5 YEAR YoY FUNDING GROWTH	35.3%
5 YEAR YoY DEAL GROWTH	40.1%

Source: *CB Insights*

DUBLIN AT A GLANCE

Power Breakfast

- Food Game
- KC Peaches
- Slattery's

Coffee Shops to Start Up In

- 3Fe
- Coffee Angel
- Joe's Coffee
- Kaph
- The Fumbally

Best Bar

- 57 The Headline
- Octagon Bar
- The Palace Bar
- Vintage Cocktail Club

"Must Be At" Local Events

- Electric Picnic
- Samhain Festival of Fire
- St. Patrick's Day Parade and Festival

In the Know

"I'm on the lash" translates to "I'm out drinking."

THE TOP

ANGELS

- Eamon Leonard
- Jack Leeney
- Kevin Abosch
- Mark Cummins
- Raj Ramanandi

VCs

- ACT Venture Capital
- AIB Seed Capital Fund
- Delta Partners
- Enterprise Equity
- Kernel Capital
- Oyster Technology Investments

ACCELERATORS

- NDRC LaunchPad
- New Frontiers

ENTREPRENEURSHIP EVENTS

- Dubstarts
- Google for Entrepreneurs Week
- OpenCoffee Club Ireland
- Startup Weekend
- ThousandSeeds

Q4'15 - Q3'16 TOP FUNDING ROUNDS

- eir—$255M
- Hibernia Networks—$165M
- Interum Therapeutics—$40M
- Mainstay Medical—$34M
- Clavis Insight—$20M
- GC Aesthetics—$20M

29

DUBAI and MENA

Almost overnight, Dubai has become one of the world's elite cities, benefitting from its central location and state-of-the-art infrastructure. It has the world's best airport and airline, arguably: Emirates. (We heard one global traveler say, "When you die, you want to go to Emirates airlines.") The cosmopolitan city is the fastest growing large city in the world, with 20 percent of all cranes operating on the planet found in Dubai and 83 percent of its population made up of immigrants. Yes, it has the tallest building in the world, Burj Khalifa, but it is also incredibly safe and has no income tax. We know a good deal when we see one.

VENTURE ACTIVITY Q4'15 - Q3'16

VC FUNDING	$611.1M
DEALS	28
EXITS	7
5 YEAR YoY FUNDING GROWTH	38.8%
5 YEAR YoY DEAL GROWTH	41.1%

Source: *CB Insights*

DUBAI AND MENA AT A GLANCE

Power Breakfast

Café Bateel
Shangri La Hotel
Urban Bistro

Coffee Shops to Start Up In

RAW Coffee
Spill the Bean

Best Bar

At.mosphere Burj Khalifa
The Jetty Lounge
Zuma

Best Closing Dinner

Le Petite Maison

"Must Be At" Local Events

Dubai Film Festival (DIFF)
Dubai Summer Surprises (DSS)
UAE Desert Challenge
World Expo 2020

In the Know

There are ATMs in Dubai that dispense gold bars.

THE TOP

ANGELS

Dr. Adel Al Sayed
Fadi Ghandour
Joi Ito
Mohammed Seyfi
Sam Hassan
Samih Toukan
Tammer Qaddumi

VCs

BECO Capital
Jabbar Internet Group
iMENA Ventures
LEAP
Middle East Venture Partners
STC Ventures
Wamda Capital

ACCELERATORS

Dubai Silicon Oasis
Flat6Labs
i360
In5

ENTREPRENEURSHIP EVENTS

ArabNet Digital Summit
Global Entrepreneurship Summit
Global Innovation Summit
SME World Summit
TiE Dubai

Q4'15 - Q3'16 TOP FUNDING ROUNDS

Souq.com—$275M
Wadi—$67M
Careem—$60M
Bayut—$20M
Propertyfinder—$20M

30

PITTSBURGH

The Steel City has long lost its mills but benefits from its timeless three rivers coming together and its world-class academic institutions, Carnegie Mellon and Pitt. The startup scene is fueled by Rolling Rock Beer and united by its passion for the Steelers. Pittsburghese, including the words "warsh" and "yinz," is the language only real Pittsburghers speak and can understand. Another key Pittburgh word is the f-bomb, which can be used in every conceivable way—as a noun, adjective, or adverb. You could say, for example, "The Steel City has 446 f-ing bridges, more than any other city in the world."

VENTURE ACTIVITY Q4'15 - Q3'16

VC FUNDING	$436.1M
DEALS	173
EXITS	24
5 YEAR YoY FUNDING GROWTH	8.0%
5 YEAR YoY DEAL GROWTH	10.9%

Source: *CB Insights*

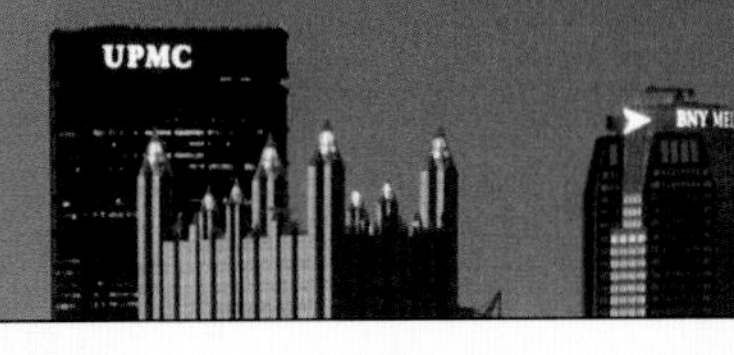

PITTSBURGH AT A GLANCE

Power Breakfast
- BarMarco
- Coco Café
- Pamela's Waffles
- Square Café

Coffee Shops to Start Up In
- 21st Street Coffee
- Big Dog
- Coffee Tree Roasters
- Commonplace Coffee
- Espresso a Mano
- Tazza D'oro
- Zeke's

Best Beer
- Rolling Rock Beer

Best Bar
- Brillo Box
- Duke's (Old Pro East)
- Grappena
- Houghts
- Industry Public House

"Must Be At" Local Events
- Asian American Film Festival
- Jam on Walnut
- Light Up Night
- Silk Screen
- Three Rivers Arts Festival
- Three Rivers Regatta
- VIA Festival

In the Know

Silicon Valley legend Bill Campbell grew up in Homestead, where half of the town is named after him and his brother.

THE TOP

ANGELS
- Audrey Russo
- Bong Koh
- Eric Silver
- Matt Newton
- Sunil Wadhwani
- William Guttman

VCs
- Adams Capital Management
- Birchmere Ventures
- BlueTree Allied Angels
- Coal Hill Ventures
- Draper Triangle Ventures
- Innovation Works
- Mellon Ventures

ACCELERATORS
- AlphaLab
- AlphaLab Gear
- Ascender
- Idea Foundry
- Starbot

ENTREPRENEURSHIP EVENTS
- Entrepreneur's Growth Conference
- Launch CMU
- Network After Work
- Pgh Tech Meetup
- Startup Weekend
- Tech50
- Thrival Festival
- UnStuck PHG

Q4'15 - Q3'16 TOP FUNDING ROUNDS
- Everest Infrastructure Partners—$100M
- Cohera Medical—$50M
- Panopto—$42.8M
- 360Fly—$40M
- Aquion Energy—$33.2M

31

MINNEAPOLIS

With a rich history of innovators, going back to companies like 3M, Control Data, and Medtronic, the Mini Apple benefits from excellent public education, a strong work ethic, and six months of the year where working is a better alternative than freezing to death. It's also known as the land of blond hair, blue eyes, and blue ears, no coincidence, given its immigrant linkage to Scandanavia and its sister city, Stockholm. It's also the medical innovation mecca of the world with the Mayo Clinic, Univeristy of Minnesota, Medical Alley and thousands of startups. With nineteen Fortune 500 companies, one Mall of America (the size of seventy-eight football fields), and sixty-nine blocks of Skyway connected over the city, the so-called Twin Cities always have something new to offer.

VENTURE ACTIVITY Q4'15 - Q3'16

VC FUNDING	427.7M
DEALS	66
EXITS	51
5 YEAR YoY FUNDING GROWTH	23.3%
5 YEAR YoY DEAL GROWTH	15.6%

Source: *CB Insights*

MINNEAPOLIS AT A GLANCE

Power Breakfast
- Edina Grill
- Good Day Café
- Original Pancake House

Coffee Shops to Start Up In
- Dunn Bros.
- Spyhouse Coffee

Best Bar
- Freehouse
- Lord Fletchers
- The Suburban

Local Beer
- Summit
- Surly Brewing Co.

"Must Be At" Local Events
- Basilica Block Party
- Guthrie Theater
- Minnesota State Fair
- Winter Carnival

In the Know

While Bob Dylan, a Minnesota native, may have won a Nobel Prize in Literature, Prince is the king of the Minneapolis music.

THE TOP

ANGELS
- David Frauenshuh
- David Russick
- Dick Perkins
- Henry Cousineau
- Jeff O'Dell
- Vance Opperman

VCs
- Arthur Ventures
- Matchstick Ventures
- Omphalos Venture Partners
- Rally Ventures
- Split Rock Partners
- StarTec Investments

ACCELERATORS
- gener8tor
- GoKart Labs
- Treehouse Health

ENTREPRENEURSHIP EVENTS
- Beta.mn
- Entrepreneur Kickoff
- Medical Alley Association
- Minnebar
- MN Cup
- MobCon

Q4'15 - Q3'16 TOP FUNDING ROUNDS
- Ceridian—$150M
- CVRx—$93M
- Code42—$85M
- Bright Health—$80M
- Siteimprove—$55M

32

MONTREAL

Located on an island in the middle of the Saint Lawrence River, Montreal has been a longtime marketplace for trading goods. A European city in North America, it is now a hub of entertainment startups, spoiling entrepreneurs with a uniquely Euro chic vibe...*trés bien*. Unlike what is commonly believed, Montrealers use the underground tunnels (twenty miles of it) mostly in the heat of summer rather than the cold of winter! Another sweet part about being an entrepreneur in Montreal: 85% of the world's maple syrup comes from Quebec.

VENTURE ACTIVITY Q4'15 - Q3'16

VC FUNDING	$442.1M
DEALS	76
EXITS	32
5 YEAR YoY FUNDING GROWTH	32.5%
5 YEAR YoY DEAL GROWTH	21.3%

Source: *CB Insights*

MONTREAL **AT A GLANCE**

Power Breakfast
Thé Mon Café

Coffee Shops to Start Up In
Aunja
Dispatch Coffee
Presse Café

Local Beer
Dieu du Ciel Péché Mortel

"Must Be At" Local Events
Igloofest
International Jazz Festival
Just for Laughs Comedy Festival

In the Know
Montreal's predominant religion is hockey.

THE TOP

ANGELS
Alan MacIntosh
Chris Arsenault
Eric Martineau-Fortin
Jonathan Nelson
Ty Danco

VCs
Chrysalix
iNovia Capital
Novacap
Real Ventures
Rho Ventures
VanEdge Capital
Yaletown

ACCELERATORS
Bolidea
FounderFuel
TandemLaunch

ENTREPRENEURSHIP EVENTS
BitNorth
Ignite
Montreal Newtech Meetup
Montreal Startup Festival
Startup Camp

Q4'15 - Q3'16 TOP FUNDING ROUNDS
Triotech Amusement—$80M
Blockstream—$55M
Coalision - Lole—$21M
Hooper—$16M
TandemLaunch—$12M

33

PHILADELPHIA

The city that Ben Franklin built is not only the birthplace of the United States, but the city where underdogs are embraced...which is perfect for startups. University City is the world's largest research park, where 17 academic institutions like Temple, Penn, Drexel, and many others congregate around entrepreneurs. The City of Brotherly Love is home to the country's first zoo, hospital, and medical school. And one out of every six doctors in the country is trained in Philly. Fortunately, they can't cure the entrepreneurial bug.

VENTURE ACTIVITY Q4'15 - Q3'16

VC FUNDING	$536.2M
DEALS	91
EXITS	42
5 YEAR YoY FUNDING GROWTH	18.2%
5 YEAR YoY DEAL GROWTH	12.7%

Source: *CB Insights*

PHILADELPHIA AT A GLANCE

Power Breakfast
- La Colombe
- La Pain Quotidien
- Minellas
- Nineteen
- PARC
- Radnor Hotel

Coffee Shops to Start Up In
- Elixr
- HubBub
- Joe
- Lovers & Madmen
- OCF

Local Beer/Shot
- Blue Coat Gin
- Dad's Hat Whiskey
- Stateside Vodka
- Tröegs
- Yuengling

Best Bar
- Union League

"Must Be At" Local Events
- Made in America
- Mummers Parade (New Years)
- XPoNential Music Festival

In the Know

All venture roads in Philly lead to Safeguard Scientific's founder Warren V. "Pete" Musser.

THE TOP

ANGELS
- David Adelman
- Doug Alexander
- Ira Lubert
- John Loftus
- John Ryan
- Josh Kopelman
- Michael Aronson
- Michael Rubin
- Mike Carter
- Pete Musser
- Sashi Reddi
- Walter Buckley

VCs
- Actua
- Bullpen Capital
- Comcast Ventures
- First Round Capital
- LLR
- MentorTech Ventures
- Milestone Partners
- NewSpring Capital
- Osage Ventures
- Rittenhouse Ventures
- Safeguard Scientific

ACCELERATORS
- Dreamit
- Drexel
- PSL Accelerator
- Quorum's Digital Health Science Center
- Tiger Labs
- UPenn

ENTREPRENEURSHIP EVENTS
- Code for Philly
- PennApps
- Philly Techweek
- Quorum
- SBN

Q4'15 - Q3'16 TOP FUNDING ROUNDS
- LibertySBF—$75M
- Accolade—$71.1M
- InstaMed—$50M
- Aprecia Pharmaceuticals—$30M
- Curalate—$27.5M

34

VANCOUVER

The birthplace of the viral business Slack, Vancouver's startup ecosystem is propelled by cheaper talent, an excellent support infrastructure, and government programs to help startups minimize cost. It's hard to imagine a more beautiful place in the world that lends itself to aspiring entrepreneurs, eh? In fact, with its ubiquitous yoga, fusion food, and parks, it is no wonder that Lululemon started here. Among large Canadian cities, Vancouver has the highest concentration of artists. Not surprisingly, the Occupy movement started here.

VENTURE ACTIVITY Q4'15 - Q3'16

VC FUNDING	$814.9M
DEALS	111
EXITS	33
5 YEAR YoY FUNDING GROWTH	38.4%
5 YEAR YoY DEAL GROWTH	19.3%

Source: *CB Insights*

VANCOUVER **AT A GLANCE**

Power Breakfast
- Cactus Club Café
- De Dutch
- The End

Coffee Shops to Start Up In
- Elysian
- Matchstick
- Timbertrain

Local Beer
- Citradelic Single Hop Citra IPA

"Must Be At" Local Events
- BC Highland Games and Scottish Festival
- Squamish Valley Music Festival
- TED

In the Know

Before starting out, check out the presentation from Ryan Holmes (CEO of Hootsuite) on why Vancouver is a great startup city including traits such as "engineering talent", "financial support", and more importantly "better beer & bacon".

THE TOP

ANGELS
- Amos Michelson
- Lance Tracey
- Markus Frind
- Mike Volker
- Norm Francis
- Ryan Holmes
- Shafin Tejani

VCs
- BDC Capital
- Lyra Growth Parters
- Stanley Park Ventures
- Vanedge Capital
- Ventures West Capital
- VersionOne

ACCELERATORS
- Invoke
- Launch Academy
- VentureLabs

ENTREPRENEURSHIP EVENTS
- Beyond Pink
- Grow Conference
- Launch @ Grow
- Lean Startup Vancouver
- Traction Conference
- Vancouver Startup Week

Q4'15 - Q3'16 TOP FUNDING ROUNDS
- Indochino—$30M
- Cymax—$25M
- Allocadia—$16.5M
- Bench—$16M
- ESSA—$15M
- Lendful—$15M
- Trulioo—$15M

35

SALT LAKE CITY

In the past two years, Utah has seen the most economic growth in the country, and is home to many "soonicorns." The Salt Lake City to Provo corridor has been a fertile environment for technology startups. While alcohol has proven to be an essential ingredient for many startups in parts of the Global Silicon Valley, channeling the prohibition of spirits has proven to benefit the entrepreneurs here. But despite its dry reputation, there are more bars than you'd think—118 and counting. There are also more non-Mormons than Mormons, and SLC has one of the largest LGBT communities in the United States.

VENTURE ACTIVITY Q4'15 - Q3'16

VC FUNDING	$364.4M
DEALS	47
EXITS	25
5 YEAR YoY FUNDING GROWTH	22.6%
5 YEAR YoY DEAL GROWTH	4.3%

Source: *CB Insights*

SALT LAKE CITY **AT A GLANCE**

Power Breakfast

Copper Onion
Pago
Park Café
Pig & a Jelly Jar

Coffee Shops to Start Up In

Coffee Connection
Park City Roasters
The Rose Establishment

Local Bar

Bar X
Beer Bar
High West Distillery

"Must Be At" Local Events

Sandy Balloon Festival
Sundance Film Festival
Twilight Concerts

In the Know

The Mormon Mafia is real. While not as deathly, it is certainly as potent. Also, the Hotel Monaco is one of the best hotels to get shut-eye in SLC.

THE TOP

ANGELS

Ben Capell
Elliott Bisnow
Joel Peterson
Josh James
Mike Levinthal
Nobu Mutaguchi

VCs

EPIC Ventures
Kickstart Seed Fund
Peak Ventures
Pelion Venture Partners
Signal Peak
Sorenson Capital
University Venture Fund
Wasatch Advisors

ACCELERATORS

BioInnovators Gateway
Boom Startup
Church & State
Impact Hub
Lassonde Institute
ThinkAtomic

ENTREPRENEURSHIP EVENTS

BrainShare
INTERFACE
RootsTech
StartFEST
Tech Security Conference

Q4'15 - Q3'16 TOP FUNDING ROUNDS

StorageCraft—$187M
Domo—$130M
Vivint—$100M
Health Catalyst—$70M
Lucid Software—$36M

36

HONG KONG

The business of Hong Kong is business. The sport of Hong Kong is business. The government of Hong Kong is business. A city with beautiful views, huge skyscrapers, and world-class hotels, Hong Kong is an economic animal that revolves around one word...business. In the car business, Hong Kong has more Rolls Royces per person than any other city in the world. And in the drinking business, there are no restrictions on selling alcohol to minors in Hong Kong. Wondering if those Thiel fellas are reading this.

VENTURE ACTIVITY Q4'15 - Q3'16

VC FUNDING	$635.0M
DEALS	74
EXITS	34
5 YEAR YoY FUNDING GROWTH	0%
5 YEAR YoY DEAL GROWTH	35.8%

Source: *CB Insights*

HONG KONG **AT A GLANCE**

Power Breakfast
- Classified
- Ethos
- Feast

Coffee Shops to Start Up In
- Coffee Academics
- Cupping Room
- Hazel & Hershey

Local Beer
- HK Dragon's Back Pale
- Young Master

"Must Be At" Local Events
- Clockenflap
- Hong Kong Rugby Sevens
- Horse racing at the Hong Kong Jockey Club
- Pecha Kucha Hong Kong

In the Know

Hong Kong's subway system is managed entirely by AI with a 99.9 percent on-time rate. Humans need not apply.

THE TOP

ANGELS
- Chipper Boulas
- Eric Kwan
- Fritz Demopoulos
- Li Ka-shing
- Robin Chan
- Yat Siu

VCs
- Arbor Ventures
- Cherubic Ventures
- Fresco Capital
- Horizons Ventures
- Sailing Capital
- Vectr Ventures

ACCELERATORS
- Blueprint
- Brinc
- CoCoon
- Cyberport
- Nest

ENTREPRENEURSHIP EVENTS
- CoCoon Pitch
- Hong Kong Launch48
- Nest Pitch Day
- Startup BootCamp
- Startup Grind
- Web Wednesday

Q4'15 - Q3'16 TOP FUNDING ROUNDS
- Baidu Waimai—$400M
- Tink Labs—$125M
- Travelzen—$92.5M
- Orange Sky Golden Harvest—$61.5M
- Asia Clean Capital—$40M

37

ANN ARBOR and DETROIT

Detroit is going through a renaissance, much like what New Orleans went through after Hurricane Katrina. Talented, passionate, driven people are flocking to Detroit to rebuild the city back to the land of innovation that it was during the days of Henry Ford. Techies have Detroit to thank for being the birthplace of techno. And remember that during Prohibition, 70 percent of all illegal alcohol that entered the United States came through Detroit. How's that for an entrepreneurial spirit?

AGGREGATE VENTURE ACTIVITY Q4'15 - Q3'16

VC FUNDING	$156.5M
DEALS	74
EXITS	15
5 YEAR YoY FUNDING GROWTH	11.4%
5 YEAR YoY DEAL GROWTH	15.1%

Source: *CB Insights*

ANN ARBOR AND DETROIT AT A GLANCE

Power Breakfast

- Cafe Zola
- Sava's
- Señor Lopez
- Zingerman's

Coffee Shops to Start Up In

- Comet Coffee
- Mighty Good
- Sweetwaters

Local Beer

- Brune
- Final Absolution
- Gulo Gulo

"Must Be At" Local Events

- Ann Arbor Summer Festival
- Detroit Downtown Hoedown
- Movement Electronic Music Festival

In the Know

In Ann Arbor, knowing Jim Harbaugh is more important than knowing the Pope.

THE TOP

ANGELS

- Dave Hartmann
- Roger Newton
- Skip Simms
- Walter Young

VCs

- Arboretum Ventures
- Beringea
- Detroit Venture Partners
- GM Ventures
- Plymouth Ventures
- RPM Ventures

ACCELERATORS

- Desai
- Spark Central
- TechArb
- Techstars Detroit
- TechTown
- UM Venture Accelerator

ENTREPRENEURSHIP EVENTS

- Accelerate Mighican
- Detroit Startup Weekend
- Entrepreneurs Engage
- Grow Detroit
- MHacks
- Michigan Growth Capital Symposium
- TechWeek

Q4'15 - Q3'16 TOP FUNDING ROUNDS

- Protean Electric—$70M
- Millendo Therapeutics—$62M
- Axios Mobile Assets—$14M
- Strata Oncology—$12M
- RetroSense Therapeutics—$6M

38

TOKYO

The Land of the Rising Sun is one of the cleanest, classiest and most organized cities in the world, with denizens that conduct business with not a hair out of place. When in Tokyo, be sure to take note of the traditional Japanese customs...one stray boot, could mean that you get the boot. Tokyo redefines "ramen profitability" and challenges founders to live a life in even less apartment space than in San Francisco. Entrepreneurs and investors look forward to making it big so they can enjoy the cuisine of a city with the most Michelin stars in the world.

VENTURE ACTIVITY Q4'15 - Q3'16

VC FUNDING	$511.0M
DEALS	97
EXITS	32
5 YEAR YoY FUNDING GROWTH	57.9%
5 YEAR YoY DEAL GROWTH	25.6%

Source: *CB Insights*

TOKYO AT A GLANCE

Power Breakfast
French Kitchen
K'Shiki
Les Saisons

Coffee Shops to Start Up In
Little Nap Coffee Stand
No8 Bear
Omotesando Koffee
Sarutahiko
Steamer Coffee

Best Bar
BAR HIGH FIVE
Shampoo
Shot Bar

Local Beer
Baird Beer Harajuku
Popeye
Saporo
Yona Yona Beer Kitchen

"Must Be At" Local Events
Fuji Rock
Pecha-Kucha
T-Site Daikanyama
Ultra Japan Music Festival

In the Know
Tokyokkos love to drink. Being able to handle your sake is as important as building an MVP, minimally viable product.

THE TOP

ANGELS
Amitt Mahajan
Eric Kwan
Joi Ito
Justin Waldron
Kazuya Minami

VCs
DCM
Globis
JAFCO
SBI Holdings
SoftBank

ACCELERATORS
DOCOMO Innovation Village
KDDI Mugen Labo
Movida
Open Network Lab
Samurai Incubate

ENTREPRENEURSHIP EVENTS
Justa
Mobile Monday Tokyo
Slush
Startup Weekend Tokyo
Tech in Asia
The Bridge

Q4'15 - Q3'16 TOP FUNDING ROUNDS
App-CM—$140M
SmartNews—$38M
BizReach—$32.9M
bitFlyer—$27M
Midokura—$20.4M

39

MADRID

With sangrias and siestas, Madrid historically was not the greatest place to do business. Now entrepreneurs have gone mad for Madrid, whose startup community gets big *besos* from the Spanish government. Students feel bullish about Madrid as it is the hot spot for university education in Europe with some of the oldest and most prestigious programs. The tide is rising for startups in Madrid as Google recently opened an innovation campus there, only its third one in Europe. While the financial crisis has rocked the country, the silver lining is engineers will "work for food" as more people venture out on their own.

VENTURE ACTIVITY Q4'15 - Q3'16

VC FUNDING	$354.8M
DEALS	58
EXITS	30
5 YEAR YoY FUNDING GROWTH	108.6%
5 YEAR YoY DEAL GROWTH	45.2%

Source: *CB Insights*

MADRID AT A GLANCE

Power Breakfast
- La Pepa Chic
- Mur Café

Coffee Shops to Start Up In
- Café del Círculo de Bellas Artes
- Google Campus Café

Best Bar
- Cervezas La Virgen
- Mahou

"Must Be At" Local Events
- Carnival
- FITUR
- Madrid Fashion Week
- Madrid Fusion

In the Know

The expression for "Cheers" is literally "Up, down, to the center, inside." Just be careful not to say that in the wrong context to a stranger you just met.

THE TOP

ANGELS
- François Derbaix
- Iñaki Arrola
- Iñaki Berenguer
- Yago Arbeloa

VCs
- Amerigo
- Cabiedes and Partners
- Fundación José Manuel Entrecanales
- Inveready
- Kibo Ventures

ACCELERATORS
- Lanzadera
- Open Future
- Seedrocket
- Tetuan Valley
- Wayra

ENTREPRENEURSHIP EVENTS
- Campus Party
- Salon MiEmpresa
- Smart Money
- South Summit
- xSpain

Q4'15 - Q3'16 TOP FUNDING ROUNDS
- Cabify—$120M
- Jobandtalent—$42M
- Fever—$12M
- Plant Response Biotech—$6M
- CloudApp—$2M

40

PORTLAND

Portland is where the California hippies moved when San Francisco went from chill to zill (as in zillionaire). Cleaner than the grungier Seattle, rainy Portland attracts the creative, free-spirited entrepreneurs with great microbreweries, roses, good food, and a welcoming environment. Finally, thank god for modernity—over a hundred years ago, Portland was considered the most dangerous city in the states to go out for a drink. Not because you might end up dead, but shanghaied on a boat to another country.

VENTURE ACTIVITY Q4'15 - Q3'16

VC FUNDING	$370.9M
DEALS	78
EXITS	30
5 YEAR YoY FUNDING GROWTH	-12.4%
5 YEAR YoY DEAL GROWTH	-5.6%

Source: *CB Insights*

PORTLAND **AT A GLANCE**

Power Breakfast
- Daily Café
- Little T American Baker

Coffee Shops to Start Up In
- Barista
- Stumptown
- Water Avenue Coffee

Local Beer
- Bridgeport IPA

Best Wine
- Chapter 24

"Must Be At" Local Events
- Bite of Oregon
- Oregon Brewers Festival
- Portland Rose Festival

In the Know

This is where you will find the real Duck Dynasty, and Phil Knight is the patriarch.

THE TOP

ANGELS
- Adam Greene
- Chris Logan
- Eric Doebele
- Nitin Khanna
- Nitin Rai

VCs
- 3x5 Special Opportunity Fund
- Oregon Angel Fund
- Portland Seed Fund
- Seven Peaks
- TiE
- Voyager

ACCELERATORS
- Forge
- Portland Incubator Experiment
- Portland Seed Fund
- PSU Business Accelerator

ENTREPRENEURSHIP EVENTS
- Angel Oregon
- Next Portland
- OEN PubTalk
- Portland Startup Weekend
- TiE Portland

Q4'15 - Q3'16 TOP FUNDING ROUNDS
- ID Experts—$27.5M
- Janrain—$27M
- Cloudability—$24M
- Cedexis—$22M
- Inpria Corporation—$10M

41

JAKARTA

Indonesia's 250 million people have contributed to making Jakarta one of the most vibrant startup scenes in Southeast Asia. There are as many new companies as Indonesia's 18,000 islands, including GoJek, which aims to be Indonesia's *uber* Uber. With its frenetic pace of life, 24-7 malls, and rush hour that makes LA look like a stroll in the park, Jakarta is also called the Big Durian, the equivalent of the Big Apple (NYC) in Indonesia. With the 4th largest population in the world living on an archipelago, the next billion dollar opportunity will surface in infrastructure, logistics, finance, and other "hard" problems.

VENTURE ACTIVITY Q4'15 - Q3'16

VC FUNDING	$1.2B
DEALS	76
EXITS	12
2 YEAR YoY FUNDING GROWTH	188.8%
5 YEAR YoY DEAL GROWTH	80.2%

Source: *CB Insights*

JAKARTA **AT A GLANCE**

Power Breakfast

- Cafe Authentique
- Le Quartier
- Paul

Coffee Shops to Start Up In

- Anomalie
- Common Grounds
- Giyanti
- Ombe Koffie

Best Bars

- Bats
- CJs

Local Beer

- Anker
- Bintang

"Must Be At" Local Events

- Festival Jalan Jaksa
- Jakarta Fair
- Java Jazz Festival

In the Know

@Jakarta: the best way to reach someone is by tweeting at them in the so-called Twitter capital of the world.

THE TOP

ANGELS

- Batara Eto
- Daniel Jones
- Kaspar Zhou
- Kuo-Yi Lim
- Willson Cuaca

VCs

- 500 Startups
- East Ventures
- GDP
- Monk's Hill Ventures
- Mountain Kejora Ventures
- Northstar Group

ACCELERATORS

- 500 Startups
- Binus University
- Ideabox
- Indigo

ENTREPRENEURSHIP EVENTS

- Echelon Indonesia
- Growth Hacking Asia—Indonesia
- Ideafest
- Startup Asia
- Startup Weekend Indonesia

Q4'15 - Q3'16 TOP FUNDING ROUNDS

- Go-Jek—$550M
- Tokopedia—$147M
- Elevenia—$50M
- Bhinneka—$22M
- Orami—$15M

DRAPER DYNASTY

If there is a VC family dynasty, it is unquestionably the Drapers. Starting with General William Henry Draper, who founded the West Coast's first venture capital firm, Draper, Gaither and Anderson, in 1958, there have been four generations of Drapers who have entered the family business. Baidu, Tesla, and Skype are but a few of the gamechangers the Drapers have backed. Not only investors in innovation but entrepreneurs as well, the Drapers have started their own VC network, a collection of online entrepreneurship classes, and an accelerator focused on Bitcoin and VR startups.

The Draper Family picure (L to R): Billy Draper, Adam Draper, Bill Draper, Tim Draper, Jesse Draper

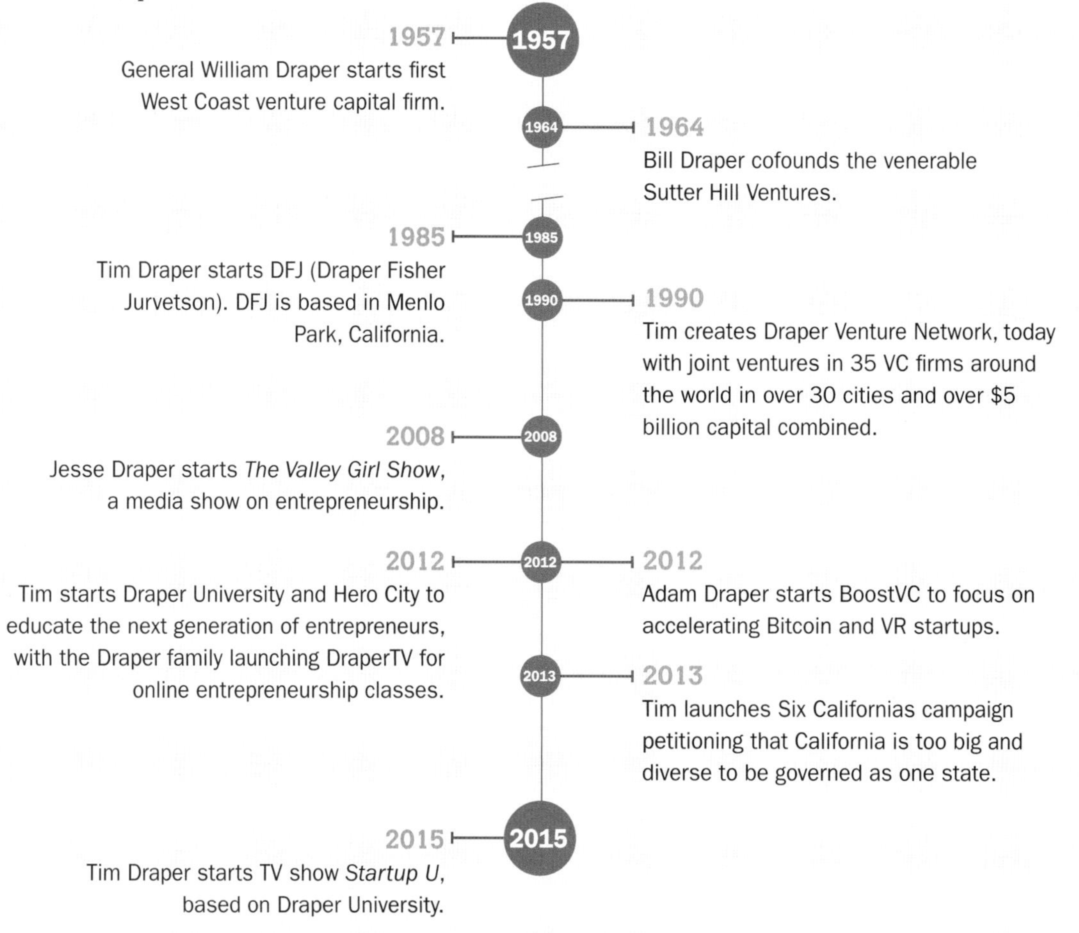

The SILICON VALLEY Job Board

Like many things in Silicon Valley, job titles are in their own Reality Distortion Field. Here are a few of the most unique job titles at top companies, as compared to what they actually mean in real life.

Sherpa Ethnic group from Nepal; one of the elite mountaineers of the Himalayas.

Innovation Sherpa: William Bunce—tasked with leading innovation at Microsoft; studied innovation at a culinary school.

Prophet Individual claiming to have been contacted by the divine who delivers newfound knowledge about the future.

Digital Prophet: David Shing, AOL, tasked with making predictions about the future for AOL; his hair is so big because it's full of secrets.

Curator A content specialist charged with the task of maintaining an institution's collection of heritage goods.

Chief Curator: Michael Moskowitz, eBay—Michael decides which used iPod or gently used fake Prada bag to feature on the front page.

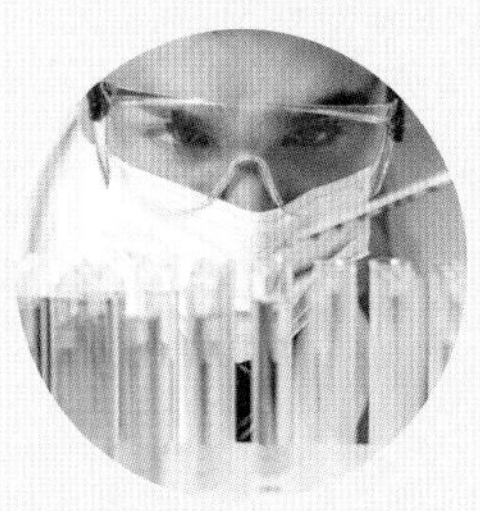

Researcher Highly educated individual who aims to study the world's most complex problems, like spontaneous generation.

Nathan Juergenson, researcher, Snapchat, student studying the complexities of disappearing digital photos... spontaneous elimination?

Philosopher A person who offers theories on profound questions in ethics and logic.

Damon Horowitz, in-house philosopher, waxes poetic about the boundaries of technology and humanities. Also moonlights as Google's director of engineering.

Genius Someone who claims to be Albert Einstein-ian in a certain field of study...and usually is.

Genius (Apple): Someone who fixes your computer... by turning it off and on.

Galactic Viceroy of Research Excellence

Real Job Title: Let's face it. There's no Galactic Viceroy of ANYTHING on this planet.

James Mickens (Microsoft), self-proclaimed title; but in reality the guy from *The Martian* was able to grow potatoes on Mars, so he's the REAL Galactic Viceroy of Research Excellence.

42

SYDNEY

Startups need go no further than Sydney in their Australian entrepreneurial walkabout, which is home to more than 60% of the startups on the continent. Beauty and beer are two of the features that have brought talent to Sydney, including Atlassian, which IPOd with the ticker TEAM. Learning some hip sayings in Chinese and Arabic will go a long way in this cosmopolitan city where both languages are immensely popular. Sydney is the capital of the Australian state of New South Wales, where the government provides an innovation concierge service to help startups best work with the government...that's rad.

VENTURE ACTIVITY Q4'15 - Q3'16

VC FUNDING	$26.2M
DEALS	6
EXITS	1
5 YEAR YoY FUNDING GROWTH	77.2%
5 YEAR YoY DEAL GROWTH	24.6%

Source: *CB Insights*

SYDNEY **AT A GLANCE**

Power Breakfast
- Bowery Lane
- Café Venti
- Foodcraft

Coffee Shops to Start Up In
- Bay Ten Espresso
- Nook Urban
- The Grounds

Best Bar
- Sky Terrace

Local Beer
- Bloody Wit
- Three Sheets

"Must Be At" Local Events
- Carols by the Tree
- St. George Open Air Cinema
- Sydney Festival

In the Know

Follow the money. Sydney is home to 65 percent of Australia's finance industry. Did someone say fintech?

THE TOP

ANGELS
- James Packer
- Mike Cannon-Brooks
- Steve Baxter

VCs
- Artesian Capital
- Blackbird Ventures
- H2 Ventures
- Sapien Ventures
- Tank Stream Ventures

ACCELERATORS
- Fishburners
- muru-D
- Stone & Chalk
- the HUB
- Tyro FintechHub

ENTREPRENEURSHIP EVENTS
- ad:tech
- CeBIT
- Microsoft Ignite
- Spark Festival
- Tech.co Celebrate
- Tech23

Q4'15 - Q3'16 TOP FUNDING ROUNDS
- Tyro Payments—$72M
- Bigcommerce—$30M
- Medical Channel—$25M
- Airtasker—$22M
- Shoes of Prey—$15.5M

43

KUALA LUMPUR

No longer in the shadow of its star sister, Singapore, Kuala Lumpur is making a name with its great startups such as GrabTaxi. Taking a page from across the Strait of Johore, the government's act to support startups has fueled a robust entrepreneurial community. Kuala Lumpur is the retail hub of Malaysia and boasts a total of sixty-six shopping malls. Knowing a big opportunity when they see one, entrepreneurs find new ways to innovate commerce, retail, and customer experience.

VENTURE ACTIVITY Q4'15 - Q3'16

VC FUNDING	$132.6M
DEALS	32
EXITS	10
2 YEAR YoY FUNDING GROWTH	264.9%
2 YEAR YoY DEAL GROWTH	146.2%

Source: *CB Insights*

KUALA LUMPUR **AT A GLANCE**

Power Breakfast
- Loveme Seven Days Café
- Take Eat Easy

Coffee Shops to Start Up In
- Butter + Beans
- Doiffee
- Shephard Star

Best Bar
- Fuego

Local Beer
- Taps Beer Bar

"Must Be At" Local Events
- MotoGP Malaysia
- Putrajaya International Hot Air Balloon Fiesta
- UrbanScapes Kuala Lumpur

In the Know

Kuala Lumpur is fondly known as KL, and its residents are known as KLites.

THE TOP

ANGELS
- Dave McClure
- Eddie Lee
- James Chan
- Kuo Yi Lim
- Vinnie Lauria

VCs
- 500 Startups
- Asia Venture Group
- Cradle Fund
- East Ventures
- Jungle Ventures

ACCELERATORS
- 1337
- Bootstrap Asia
- MAD
- Tune Labs

ENTREPRENEURSHIP EVENTS
- HackerNest Tech Socials
- Startup Grind
- SWELL
- Technopreneur Open Day

Q4'15 - Q3'16 TOP FUNDING ROUNDS
- iFlix—$45M
- iCar Asia—$13.2M
- AppShack—$4M
- Kaodim—$4M
- Ebizu—$3M
- Jirnexu—$3M

HELSINKI

Beginning with Nokia, the Abraham of the Finnish startup scene, Helsinki's entrepreneurial community has been fueled by a strong education system. Hot game companies such as Rovio and Supercell, and a long winter, which provide a winning combination for lasting innovation. Super hot innovation conference Slush attracts over 20,000 attendees to Helsinki in the middle of the winter. From using a squirrel as their symbol, to having the Wife Carrying Championships (exactly what it sounds like), to throwing cell phones as competition, the Finns have a wealth of self-amusement activities for when they aren't building companies.

VENTURE ACTIVITY Q4'15 - Q3'16

VC FUNDING	$149.7M
DEALS	59
EXITS	24
5 YEAR YoY FUNDING GROWTH	104.4%
5 YEAR YoY DEAL GROWTH	63.8%

Source: *CB Insights*

HELSINKI **AT A GLANCE**

Power Breakfast

- Café TinTin Tango
- Hotel Seurahuone
- SIS Deli

Coffee Shops to Start Up In

- Good Life Coffee
- Gran Delicato
- Moko Market
- Steam Coffee

Best Bar

- Leijuva Lahna

Local Beer

- Mufloni Vaalea
- Suomenlinna Pils
- Suomenlinnan Panimo
- Teerenpeli

"Must Be At" Local Events

- April 30th
- Flow Festival
- Helsinki Festival
- Tuska

In the Know

Modest Finns are proud of being #1 in education, #1 in games and mobile, and #1 in saunas. You will soar if you know the Mighty Eagle.

THE TOP

ANGELS

- Brett Mason
- Christian Thaler-Wolski
- Jyri Engeström
- Net Jacobsson
- Peter Vesterbacka
- Risto Siilasmaa
- Sean Seton-Rogers

VCs

- Accel Partners
- Atomico
- Initial Capital
- InVenture
- Lifeline Ventures
- London Venture
- Nexit Ventures
- Open Ocean
- PROfounders Capital

ACCELERATORS

- Koppicatch
- NewCo
- Startup Sauna
- xEdu

ENTREPRENEURSHIP EVENTS

- European Business Angel
- Helsinki Opencoffee Meetup
- Lean Startup Helsinki
- Slush
- Startup Weekend Helsinki

Q4'15 - Q3'16 TOP FUNDING ROUNDS

- Wolt—$12.4M
- Jolia—$12M
- Canatu—$11.3M
- Next Games—$10M
- GreenStream Network—$4.5M

45

NASHVILLE

Southern charm and country music aren't the only things that define Tennessee's capital. Nashville has a lively startup scene, but it is considered dim when compared to the vibrant party scene. Speaking of entrepreneurship, not only did Elvis create a whole new category of music, but he also recorded over two hundred songs in Nashville. Health care and music are Nashville's jams, as the city has a dense concentration of the key players in those industries. A must visit is also the Nashville Entrepreneur Center, which is the heart of the Nashville startup community and has worked with over 450 startups since 2010.

VENTURE ACTIVITY Q4'15 - Q3'16

VC FUNDING	$229.0M
DEALS	40
EXITS	31
5 YEAR YoY FUNDING GROWTH	7.4%
5 YEAR YoY DEAL GROWTH	-5.8%

Source: *CB Insights*

NASHVILLE **AT A GLANCE**

Power Breakfast
- Marsche
- Pancake Pantry
- Pinewood Social

Coffee Shops to Start Up In
- Crema
- Edgehill Cafe
- Frothy Monkey

Local Beer/Shot
- American Born Moonshine
- Black Abby
- Corsair
- Jack Daniels
- Yazoo

"Must Be At" Local Events
- Bonnaroo
- Live on the Green Music Festival
- Tomato Art Festival

In the Know

Don't get trapped in the touristy Opryland Hotel. You might never find your way out.

THE TOP

ANGELS
- Nick Ogden
- Pat Dillingham
- Sid Chambless
- Townes Duncan
- Vic Gatto

VCs
- Claritas Capital
- Nashville Capital Network
- Richland Ventures
- Solidus Company
- TriStar Tech Ventures

ACCELERATORS
- Jumpstart Foundry
- Nashville Entrepreneur Center

ENTREPRENEURSHIP EVENTS
- Grow Nashville
- Launch Tennessee
- Startup Weekend Nashville
- WELD Nashville Events

Q4'15 - Q3'16 TOP FUNDING ROUNDS
- Silicon Ranch—$100M
- Digital Reasoning Systems—$40M
- INSBANK—$13M
- MindCare Solutions—$12.8M
- evermind—$5M
- GoNoodle—$5M

46

COPENHAGEN

The capital of Denmark, named as one of the best countries for business in the world by *Forbes*, has a cool and active startup scene. The Danes have a long history of making waves in the global tech community through the creation of major programming languages. Since its founding, Copenhagen has endured attacks from Wendish pirates, British ships, Nazi tanks, and Swedish mercenaries. These days, the main invaders are tourists: approximately nine million visited in 2014. It is now one of the happiest cities and the best places to live in the world.

VENTURE ACTIVITY Q4'15 - Q3'16

VC FUNDING	$177.5M
DEALS	51
EXITS	10
5 YEAR YoY FUNDING GROWTH	-22.6%
5 YEAR YoY DEAL GROWTH	29.5%

Source: *CB Insights*

COPENHAGEN **AT A GLANCE**

Power Breakfast
Sweet Treat
The Union Kitchen

Coffee Shops to Start Up In
Coffee Collective
Coffee First
Original Coffee

Best Bar
Mikkeller and Friends

Local Beer
Mikkeller

"Must Be At" Local Events
CPH:DOX
Distortion
Roskilde Festival

In the Know
Danes congregate at Gråbrødretorv and Nyhavn to see, be seen, eat, drink, and be merry.

THE TOP

ANGELS
David Helgason
Hampus Jakobsson
Klaus Nyengaard
Thomas Madsen-Mygdal
Tommy Andersen

VCs
Creandum
North-East Venture
Northcap Partners
Northzone
SEED Capital
Sunstone Capital

ACCELERATORS
Accelerace
Danish Tech Challenge
Startupbootcamp
Thinkubator

ENTREPRENEURSHIP EVENTS
#CPHFTW Town Halls
Coldfront
Forge Copenhagen
LeanCPH
Startup Weekend
TechBBQ

Q4'15 - Q3'16 TOP FUNDING ROUNDS
Adform—$21.5M
Planday—$14M
Peakon—$4.4M
Bownty—$3.8M
Drivr—$3.4M

47

HAMBURG

While most of the world sees Berlin as the hottest startup city in Germany, insiders claim that Hamburg is where it happens. Hamburgers have defied the odds by building a startup ecosystem in a city of the oldest opera house, old money, old ideas, and old people. You can use any one of its 2,300 bridges to navigate to Hamburg's shopping centers, such as its urban promenade Jungfernstieg. Hamburg saw more new startups in 2015 than any year previous, with FinTech a huge employer in the northern German city.

VENTURE ACTIVITY Q4'15 - Q3'16

VC FUNDING	$231.1M
DEALS	33
EXITS	18
5 YEAR YoY FUNDING GROWTH	72.3%
5 YEAR YoY DEAL GROWTH	36.4%

Source: *CB Insights*

HAMBURG AT A GLANCE

Power Breakfast
- Pauline Café and Bistro
- Savory

Coffee Shops to Start Up In
- Balzac Coffee
- Betahaus
- Daily Coffee and Food Co.
- Eclair au Café

Best Bar
- Le Lion

Local Beer/Shot
- Brewcifer
- Ratsherrn

"Must Be At" Local Events
- Christopher Street Day Parade
- Hamburg Dom
- Hamburg Harbor Birthday
- MS Dockville

In the Know

"Mors Mors" translates into "kiss my ass" but is Hamburg sports fans' friendly way of cheering for their home teams.

THE TOP

ANGELS
- Lars Hinrichs
- Nils Regge
- Philipp Schroeder
- Rolf Mathies
- Stephan Uhrenbacher
- Ulrich Hegge

VCs
- DS Investment
- e.ventures
- IFB Innovationsstarter
- iVenture Capital
- Shortcut Ventures
- Tru Venturo

ACCELERATORS
- Airbus BizLab
- Comdirect Startup Garage
- Dynport GmbH
- Greenhouse Innovation
- Hanse Ventures
- Next Media Accelerator
- New Commercial Room

ENTREPRENEURSHIP EVENTS
- Founder Academy
- Founders Talk
- Hamburg Innovation Summit
- Ideensturm
- LaborX Series
- NEUMACHER Conference
- Real Time Hamburg

Q4'15 - Q3'16 TOP FUNDING ROUNDS
- Kreditech—$103M
- FinanzCheck—$37.5M
- Deposit Solutions—$16.4M
- Exporo AG—$9.1M
- Topas Therapeutics—$4.5M

48

MOSCOW

Despite the red tape from the Reds, Moscow is where many young brainiacs call home and aim to build the next game-changing startup. Their only hindrance? See page 121, "New Jersey." Stop by the Skolkovo Innovation Center to check out the innovation city backed by the government and other power brokers. Moscow redefines angel investing, and Forbes ranks the city third in terms of the number of billionaires it has.

VENTURE ACTIVITY Q4'15 - Q3'16

VC FUNDING	$306.8M
DEALS	23
EXITS	8
5 YEAR YoY FUNDING GROWTH	-1.7%
5 YEAR YoY DEAL GROWTH	-13.7%

Source: *CB Insights*

MOSCOW AT A GLANCE

Power Breakfast
Café Michel

Coffee Shops to Start Up In
Bucer's
Cafe Artista

Offline Tinder
Soho Rooms

Local Beer
Baltika
Jigulevskoe
Klinskoye Svetloe

"Must Be At" Local Events
Contemporary Music Festival
Maslenitsa
Russia's Independence Day
The Kremlin Cup
Woman's Day

In the Know
Ask for money, get vodka.
Ask for vodka, get money twice.

THE TOP

ANGELS
Alexander Pavlov
Dmitry Falkovich
Dmitry Grishin
Edward Shenderovich
Oleg Tscheltzoff
Yuri Milner

VCs
Direct Group
DST Global
e.Ventures
Intel Capital
Kinnevik

ACCELERATORS
API Moscow
Navigator Campus
Skolkovo Innovation Center
TechPeaks
Yandex

ENTREPRENEURSHIP EVENTS
Lean Startup
Moscow Startup Founder 101
Startup Sauna
Startup Village
TechCrunch

Q4'15 - Q3'16 TOP FUNDING ROUNDS
YouDo—$6.2M
Chefmarket—$5M
OneTwoTrip—$4M
PayQR—$3.9M
Genotek—$2M

49

JOHANNESBURG

Johannesburg has more mobile phones in Africa than there are people, requiring entrepreneurs here to live, breathe, and eat "mobile first." Facebook also opened their first office in Africa here. As the largest city in the world not built on a coastline, lake, or river, Johannesburg began its life as a gold-mining town, while people flock there now for digital gold. Unbeknownst to many, South African startups raised some $54.6 million in 2015, the highest on the continent.

VENTURE ACTIVITY Q4'15 - Q3'16	
VC FUNDING	$146.6M
DEALS	17
EXITS	12
2 YEAR YoY FUNDING GROWTH	-84.4%
2 YEAR YoY DEAL GROWTH	112.5%

Source: *CB Insights*

JOHANNESBURG **AT A GLANCE**

Power Breakfast
Four Seasons
Love Food
Salvation Café

Coffee Shops to Start Up In
Bean There Cafe
Guru
Motherland

Local Beer
Castle 1895 Draught

Best Bar
Stanley Beer Yard

"Must Be At" Local Events
I Heart Joburg
Jozi Film Festival

In the Know
Known fondly by outsiders as Joburg, and by insiders as Jozi.

THE TOP

ANGELS
Charles Lorenceau
Marc Elias
Mohamed Nanabhay
Petar Soldo
Thabang Mashiloane

VCs
AngelHub Ventures
CRE Ventures
Knife Capital
Redwood Capital
Umbono Capital
Vantage Capital

ACCELERATORS
Seed Engine
Seed Institue
Tech Lab Africa

ENTREPRENEURSHIP EVENTS
AfricaCom
Mediatech Africa
Startup Grind Johannesburg
Startup Weekend Johannesburg
Tech4Africa

Q4'15 - Q3'16 TOP FUNDING ROUNDS
Snapt—$1M
Giraffe—$500K
Delvv.io—$127M
WizzPass—$120K

LAGOS

Home to Africa's first Unicorn, Africa Internet Group, Lagos is known as Africa's Big Apple. Here, entrepreneurs battle the odds, sometimes without electricity and running water, to create digital-commerce and payment startups. Startups trying to stay productive in Lagos have to navigate its many exciting distractions, including beautiful beaches, a rocking night scene, funny entertainment, and colorful owambe parties, and stave off the hunger induced from thinking about street foods such as agege, akara, boli, puff puff, abacha, agbo jedi, and much more.

VENTURE ACTIVITY Q4'15 - Q3'16

VC FUNDING	$747.4M
DEALS	36
EXITS	4
5 YEAR YoY FUNDING GROWTH	147.8%
5 YEAR YoY DEAL GROWTH	78.3%

Source: *CB Insights*

LAGOS AT A GLANCE

Power Breakfast
- Lagoon Restaurant
- Sheraton Lagos
- Terra Kulture

Coffee Shops to Start Up In
- Café Neo
- Paris Deli

Best Bar
- Quilox

Local Beer
- Orijin
- Star Lager

"Must Be At" Local Events
- Evo Festival
- Lagos Carnival
- Lagos Music Festival

In the Know

If someone tells you they want to dash, a.k.a known as a bribe, you should dash away.

THE TOP

ANGELS
- Aaron Liew
- Bunmi Akinyemiju
- Chika Nwobi
- Cornelius Frey
- Dotun Olowoporoku
- Idris Ayodeji Bello
- Opeyemi Awoyemi
- Tony Elumelu

VCs
- CRE Ventures
- EchoVC
- IFC
- Omidyar Network

ACCELERATORS
- 440NG
- Co-Creation Hub
- IDEA/Tech Launchpad
- Spark
- Wennovation Hub

ENTREPRENEURSHIP EVENTS
- Afropolitan Vibes
- DEMO Africa
- Founder2be
- Global Entrepreneurship Week
- Lagos Startup Week
- Social Media Week
- Startup Grind Lagos
- VC4Africa Meetup

Q4'15 - Q3'16 TOP FUNDING ROUNDS
- Zinox Technologies—$25M
- Landmark Africa—$20M
- Paga—$13M
- Aella Credit—$7.2M
- Travelbeta—$2M

The Global SILICON VALLEY PIONEER 250

THE GLOBAL SILICON VALLEY PIONEER 250

The businesses that generate the most spectacular returns are small companies that become big. At GSV, our objective is to identify and invest in the Stars of Tomorrow—the fastest-growing, most innovative companies in the world.

Our top-down perspective focuses on megatrends, or the technological, economic, and social forces that develop from a groundswell and eventually disrupt the status quo. We believe that understanding today's megatrends provides us with a road map to where future market opportunities are developing. Our bottom-up analysis is centered on the Four P's—People, Product, Potential, and Predictability

We have identified 250 private companies that are poised to become the next Unicorns of the world. One of the characteristics of great companies is that they are systematic and strategic in how they operate their business. Only when all Four P's are aligned can a company realize sustained long-term growth.

17zuoye — 17zuoye.com
21 Inc. — 21.co
Addepar — addepar.com
Affirm — affirm.com
Airware — airware.com
AlienVault — alienvault.com
AltSchool — altschool.com
Andela — andela.com
Anki — anki.com
App Annie — appannie.com
Aquion Energy — aquionenergy.com
Aspiration — aspiration.com
Athos — liveathos.com
August — august.com
Ayasdi — ayasdi.com
Betterment — betterment.com
BetterWorks — betterworks.com
Big Switch Networks — big-switch.com
BigBasket — bigbasket.com
BlackBuck — blackbuck.com
Blend Labs — blendlabs.com
Blockstream — blockstream.com
Bloomreach — bloomreach.com
Blue Jeans Network — bluejeans.com
Bolt Threads — boltthreads.com
Branch Metrics — branch.io
Breather — breather.com
Bright Health — brighthealthplan.com
Bromium — bromium.com
Byju's — byjus.com
Cadre — cadre.com
Canary — canary.is
Canva — canva.com
Capriza — capriza.com
CareCloud — carecloud.com
Carousell — carousell.com
Casper — casper.com
ChargePoint — chargepoint.com
Checkr — checkr.com
Circle — circle.com
CircleUp — circleup.com
Civitas Learning — civitaslearning.com
Clarifai — clarifai.com
Class Dojo — classdojo.com
ClassPass — classpass.com
Clean Power Finance — cleanpowerfinance.com
ClearStory Data — clearstorydata.com
Clever — clever.com
Clover Health — cloverhealth.com
Coinbase — coinbase.com
Collective Health — collectivehealth.com
Color Genomics — getcolor.com
Common — hicommon.com
CommonBond — commonbond.co
Compass — compass.com
Couchbase — couchbase.com
Counsyl — counsyl.com
Course Hero — coursehero.com
Coursera — coursera.org
CreativeLive — creativelive.com
Cumulus Networks — cumulus-networks.com
Dashlane — dashlane.com
Datadog — datadog.com
Dialpad — dialpad.com
Digital Asset — digitalasset.com
DigitalOcean — digitalocean.com

DNAnexus — dnanexus.com
DoorDash — doordash.com
Douyu — douyu.com
Drawbridge — drawbridge.com
DroneDeploy — dronedeploy.com
Duo Security — duosecurity.com
Duolingo — duolingo.com
Dwolla — dwolla.com
Earnest — meetearnest.com
Eero — eero.com
Ehang — ehang.com
Elastic — elastic.co
Endgame — endgame.com
Enjoy — goenjoy.com
Everfi — everfi.com
Flexport — flexport.com
FiveStars — fivestars.com
Flexport — flexport.com
ForeScout Technologies — forescout.com
Foursquare — foursquare.com
Freshdesk — freshdesk.com
Fundbox — fundbox.com
General Assembly — generalassemb.ly
Genius — genius.com
Ginkgo Bioworks — ginkgobioworks.com
Giphy — giphy.com
GitLab — gitlab.com
Go-Jek — go-jek.com
GoEuro — goeuro.com
GoFundMe — gofundme.com
Greenhouse — greenhouse.io
Grofers — grofers.com
Guardant Health — guardanthealth.com

HackerOne — hackerone.com
Handy — handy.com
Harry's — harrys.com
Hearsay Social — hearsaysocial.com
Hired — hired.com
Honor — joinhonor.com
HootSuite — hootsuite.com
HourlyNerd — hourlynerd.com
Hyperloop Technologies — hyperlooptech.com
IFTTT — ifttt.com
Impossible Foods — impossiblefoods.com
Instart Logic — instartlogic.com
Intercom — intercom.io
InVision — invisionapp.com
Ionic Security — ionicsecurity.com
Iora Health — iorahealth.com
iZettle — izettle.com
JAMF — jamfsoftware.com
Jaunt — jauntvr.com
JIBO — myjibo.com
Joya Communications — getjoya.com
Juicero — juicero.com
Kensho — kensho.com
Keybase — keybase.io
Knewton — knewton.com
Kymeta — kymetacorp.com
Leanplum — leanplum.com
LegalZoom — legalzoom.com
Lemonade — lemonade.com
LendUp — lendup.com
Lever — lever.co
Lightneer — lightneer.com

Livongo Health — livongo.com
Looker — looker.com
Luxe — luxe.com
Lytro — lytro.com
Malwarebytes — malwarebytes.org
Matterport — matterport.com
Medium — medium.com
MemSQL — memsql.com
Mesosphere — mesophere.com
Meta — metavision.com
MetroMile — metromile.com
Mixpanel — mixpanel.com
MobiKwik — mobikwik.com
Modern Meadow — modernmeadow.com
Munchery — munchery.com
Musical.ly — musical.ly
Namely — namely.com
Narrative Science — narrativescience.com
Navdy — navdy.com
NerdWallet — nerdwallet.com
NewVoiceMedia — newvoicemedia.com
NextVR — nextvr.com
Ninebot — ninebot.com
NuBank — nubank.com.br
nuTonomy — nutonomy.com
One Medical Group — onemedical.com
Onshape — onshape.com
OpenEnglish — openenglish.com
OpenGov — opengov.com
Operator — operator.com
Optimizely — optimizely.com
Orbital Insight — orbitalinsight.com

Outbrain — outbrain.com
OYO Rooms — oyorooms.com
Ozy Media — ozy.com
Patreon — patreon.com
Payoneer — payoneer.com
Peloton Interactive — pelotoncycle.com
Personal Capital — personalcapital.com
PillPack — pillpack.com
Pindrop Security — pindropsecurity.com
Pivot3 — pivot3.com
PlanetLabs — planet.com
PlanGrid — plangrid.com
PluralSight — pluralsight.com
Postmates — postmates.com
Procore Technologies — procore.com
Project Frog — projectfrog.com
Quantopian — quantopian.com
Quizlet — quizlet.com
Qumulo — qumulo.com
Raise — raise.com
Realm — realm.io
Redfin — redfin.com
Refinery29 — refinery29.com
Remind — remind.com
Revinate — revinate.com
Ring — ring.com
Ripple — ripple.com
Robinhood — robinhood.com
Rubrik — rubrik.com
Samsara — samsara.com
Sauce Labs — saucelabs.com
Science Exchange — scienceexchange.com
Scopely — scopely.com
Scribd — scribd.com
Segment — segment.com
SendGrid — sendgrid.com
Shape Security — shapesecurity.com
Shapeways — shapeways.com
Shift — shift.com
Sift Science — siftscience.com
Siluria Technologies — siluria.com
SiSense — sisense.com
Skyhigh — skyhighnetworks.com
SmartDrive Systems — smartdrive.net
SnapLogic — snaplogic.com
Snowflake Computing — snowflake.net
Spiceworks — spiceworks.com
Spire — spire.com
Spredfast — spredfast.com
Squarespace — squarespace.com
Stack Exchange — stackexchange.com
Stitch Fix — stitchfix.com
Sumo Logic — sumologic.com
Swiggy — swiggy.com
Synack — synack.com
Taboola — taboola.com
Takealot Online — takealot.com
Talend — talend.com
Tamr — tamr.com
Tastemade — tastemade.com
Teachers Pay Teachers — teacherspayteachers.com
Tealium — tealium.com
Teespring — teespring.com
The Muse — themuse.com
ThousandEyes — thousandeyes.com
Thrive Market — thrivemarket.com
Tilt — tilt.com
Toast — pos.toasttab.com
Tokopedia — tokopedia.com
Tradesy — tradesy.com
Truecaller — truecaller.com
Turbonomic — turbonomic.com
uBiome — ubiome.com
Udemy — udemy.com
Upstart — upstart.com
Upwork — upwork.com
VeloCloud — velocloud.com
VIPkid — vipkid.com.cn
vIPtela — viptela.com
Walker & Company Brands — walkerandcompany.com
Wealthfront — wealthfront.com
Welltok — welltok.com
WeTransfer — wetransfer.com
x.ai — x.ai
xAd — xad.com
Zipline — flyzipline.com
ZipRecruiter — ziprecruiter.com
Zuora — zuora.com
Zymergen — zymergen.com

The Global
SILICON VALLEY

THE UNICORNS

Despite the Initial Public Offering (IPO) market being weak for much of the past fifteen years, Venture Capitalists haven't stopped investing. In 2000, there was one Unicorn. Today, there are over 175.

This can be accounted for VCs who have invested in an average of 3,200 companies per year since 2001, including 3,709 companies in 2015 alone. We estimate that there are currently over 2,000 VC-backed private companies with a market value of $100 million or greater.

The combination of an IPO market which has been reduced by 80% since the 90s, the corresponding time from VC investment to monetization quadrupling, and the digital tracks that have been laid, which connects 3.1 billion people on Internet and 2.6 billion people on smartphones have created opportunity to go from idea to reaching tens of millions of people at the blink of an eye.

1. Uber ($62.5B)
2. ANT Financial ($60.0B)
3. Xiaomi ($45.0B)
4. Didi Chuxing ($33.7B)
5. Airbnb ($30.0B)
6. Palantir Technologies ($20.1B)
7. Snapchat ($19.3B)
8. Lufax ($18.5B)
9. Meituan-Dianping ($18.0B)
10. WeWork ($16.9B)
11. Flipkart ($15.0B)
12. SpaceX ($12.0B)
13. Pinterest ($11.0B)
14. Dropbox ($10.4B)
15. Stripe ($9.2B)
16. Spotify ($8.5B)
17. DJI ($8.0B)
18. ZhongAn ($8.0B)
19. Cainiao Logistics ($7.7B)
20. JD Finance ($7.1B)
21. Hulu ($5.8B)
22. Home Link (Lianjia) ($5.7B)
23. Lyft ($5.5B)
24. Coupang ($5.0B)
25. Ola ($5.0B)
26. Paytm ($4.8B)
27. Snapdeal ($4.8B)
28. Magic Leap ($4.5B)
29. UCAR ($4.4B)
30. Cloudera ($4.1B)
31. Yello Mobile ($4.05B)
32. Slack ($3.8B)
33. Garena ($3.8B)
34. Greensky ($3.6B)
35. Credit Karma ($3.5B)
36. LeSports ($3.3B)
37. Meizu ($3.3B)
38. Fanatics ($3.2B)
39. Delivery Hero ($3.1B)
40. BAM Tech ($3.0B)
41. DocuSign ($3.0B)
42. Ele.me ($3.0B)
43. Grab ($3.0B)
44. HelloFresh ($3.0B)
45. Ping An Good Doctor ($3.0B)
46. SoFi ($3.0B)
47. VANCL ($3.0B)
48. Wanda E-commerce ($3.0B)
49. Moderna Therapeutics ($3.0B)
50. Bloom Energy ($2.8B)
51. Pivotal ($2.8B)
52. Powa Technologies ($2.8B)
53. Oscar ($2.7B)
54. Vice Media ($2.6B)
55. Lazada Group ($2.5B)
56. Mozido ($2.4B)

57 . . . . Meitu ($2.4B)
58 .Houzz ($2.3B)
59 .Adyen ($2.3B)
60 . . . . Alibaba Pictures ($2.1B)
61 . . . . Taobao Movie ($2.1B)
62 . . . . AppDynamics ($2.1B)
63 . . . . Avant ($2.0B)
64 . . . . .Beijing Weiying Technology ($2.0B)
65 . . . . Blue Apron ($2.0B)
66 .Domo ($2.0B)
67 . . . . Firstp2p ($2.0B)
68 . . . . Gett ($2.0B)
69 . . . . GitHub ($2.0B)
70 . . . . InMobi ($2.0B)
71 . . . . Instacart ($2.0B)
72 . . . . .Intarcia Therapeutics ($2.0B)
73 . . . . . . One97 Communications ($2.0B)
74 . . . . .SurveyMonkey ($2.0B)
75 . . . . .Trendy International Group ($2.0B)
76. . . . . Zenefits ($2.0B)
77 . . . . Prosper ($1.9B)
78 . . . . Avito.ru ($1.8B)
79 . . . . NantOmics ($1.8B)
80 . . . . Sprinklr ($1.8B)
81 . . . . Tanium ($1.8B)
82 . . . . ZocDoc ($1.8B)
83 . . . . CureVac ($1.7B)
84 . . . . Honest Co. ($1.7B)
85 . . . . .Jawbone ($1.7B)
86. . . . . Deem ($1.6B)
87 . . . . Lakala ($1.6B)
88 . . . . Quanergy Systems ($1.6B)
89 . . . . Zoox ($1.6B)
90 . . . . Blippar ($1.5B)
91 . . . . BuzzFeed ($1.5B)
92 . . . . Farfetch ($1.5B)
93 . . . . .InsideSales.com ($1.5B)
94 . . . . JetSmarter ($1.5B))
95 . . . . LivingSocial ($1.5B)
96. . . . . .Mu Sigma ($1.5B)
97 . . . . MuleSoft ($1.5B)
98 . .Oxford Nanopore Technologies ($1.5B)
99 . . . . Unity Technologies ($1.5B)
100 . . . . .WeDoctor (Guahao) ($1.5B)
101 . . . . Conduit ($1.4B)
102 .Hike ($1.4B)
103 .Razr ($1.4B)
104 . . . . Koudai ($1.4B)
105 . . . . MongoDB ($1.4B)
106. . . . . 58 Daojia ($1.3B)
107 . . . . Apttus ($1.3B)
108 . . . . Auction.com ($1.3B)
109 . . . . Dada ($1.3B)
110 . . . . FanDuel ($1.3B)
111 . . . . .GoJek ($1.3B)
112 . . . . Medallia ($1.3B)
113 . . . . Red Ventures ($1.3B)
114 . . . . Thumbtack ($1.3B)
115 .Tujia ($1.3B)
116. . . . . .AppNexus ($1.2B)
117 . . . . AUTO1 Group ($1.2B)
118 . . . . Automattic ($1.2B)
119 . . . . .BlaBlaCar ($1.2B)
120 . . . . DraftKings ($1.2B)
121 . . . . Human Longevity ($1.2B)
122 . . . . Infinidat ($1.2B)
123 . . . . Klarna ($1.2B)
124 . . . . Lashou.com ($1.2B)
125 . . . . Miaopai ($1.2B)
126. . . . . Okta ($1.2B)
127 . . . . Proteus Digital Health ($1.2B)
128 . . . . Qualtrics ($1.2B)
129 . . . . Sogou ($1.2B)
130 . . . . Ten-X ($1.2B)
131 . . . . Warby Parker ($1.2B)
132 . . . . 23andMe ($1.1B)
133 . . . . .Actifio ($1.1B)
134 . . . . Adknowledge ($1.1B)
135 . . . . Anaplan ($1.1B)
136. . . . . CloudFlare ($1.1B)

137 . . . . Cylance ($1.1B)
138 . . . . .Eventbrite ($1.1B)
139 . . . . .Funding Circle ($1.1B)
140 . . . . Global Fashion Group ($1.1B)
141 . . . . MarkLogic ($1.1B)
142 . . . . MediaMath ($1.1B)
143 . . . . MindMaze ($1.1B)
144 . . . . Nextdoor ($1.1B)
145 . . . . NJOY ($1.1B)
146. . . . . ShopClues ($1.1B)
147 . . . . Taboola ($1.1B)
148 .Tango ($1.1B)
149 . . . . .TransferWise ($1.1B)
150 . . . . Vox Media ($1.1B)
151 . . . . Wifi Skeleton Key ($1.1B)
152 . . . . Zomato ($1.1B)
153 . . . . .Zscaler ($1.1B)
154 .Zuora ($1.1B)
155 . . . . .Africa Internet Group ($1.0B)
156. . . . . .Age of Learning ($1.0B)
157 . . . . AppDirect ($1.0B)
158 . . . . AVAST Software ($1.0B)
159 . . . . BeiBei ($1.0B)
160 . . . . .Boohoo.com ($1.0B)
161 . . . . Compass ($1.0B)
162 . . . . Datto ($1.0B)
163 . . . . Deliveroo ($1.0B)
164 . . . . Docker ($1.0B)
165 . . . . Douyu TV ($1.0B)
166. . . . . Evernote ($1.0B)
167 . . . . Evolution Gaming ($1.0B)
168 . . . . .Fanli ($1.0B)
169 . . . . . .ForeScout Technologies ($1.0B)
170 . . . . .Glassdoor ($1.0B)
171 . . . . Home24 ($1.0B)
172 . . . . iCarbonX ($1.0B)
173 . . . . Illumio ($1.0B)
174 . . . . ironSource ($1.0B)
175 . . . . .iTutorGroup ($1.0B)
176. . . . . lwjw ($1.0B)
177 . . . . JustFab ($1.0B)
178 .Kabam ($1.0B)
179 . . . . .Kabbage ($1.0B)
180 . . . . .Kik ($1.0B)
181 . . . . .Lamabang ($1.0B)
182 . . . . .Little Red Book ($1.0B)
183 . . . . Lookout ($1.0B)
184 . . . . Mercari ($1.0B)
185 . . . . Mime Cast ($1.0B)
186. . . . . Mogujie ($1.0B)
187 . . . . OfferUp ($1.0B)
188 . . . . .OpenDoor ($1.0B)
189 . . . . Panshi ($1.0B)
190 . . . . Procure Technologies ($1.0B)
191 . . . . . QuEST Global Services ($1.0B)
192 . . . . Quikr ($1.0B)
193 . . . . Rong360 ($1.0B)
194 . . . . Shazam ($1.0B)
195 . . . . .SimpliVity ($1.0B)
196. . . . . .SMS Assist ($1.0B)
197 . . . . Souq ($1.0B)
198 . . . . Tintri ($1.0B)
199 . . . . U51.com ($1.0B)
200 .Ubtech ($1.0B)
201 . . . . Udacity ($1.0B)
202 . . . . Wandoujia ($1.0B)
203 . . . . Wish ($1.0B)
204 . . . . Yidao Yongche ($1.0B)
205 . . . . .Zeta Interactive ($1.0B)

Source: GSV Asset Management, TechCrunch, Wall Street Journal, and CB Insights.

The Global SILICON VALLEY

ACKNOWLEDGMENTS

ACKNOWLEDGMENTS

It's been ten years since I wrote my last book, *Finding the Next Starbucks*, and like starting a second business, it's advantage to have a poor memory as not to remember the amount of pain there is to get a book written and published.

Thankfully, I had a team of supporters and contributors that helped me keep going, even when I wasn't really sure where I was headed. Namely, my wife Bonnie has been a cheerleader for many years and tough coach when I needed it. My daughters Maggie and Caroline, who keep me up on what's cool without making me feel like I'm a relic. An additional thank you to my oldest daughter Maggie for her hilarious edits and making sure I didn't take for granted that most normal people have no idea what an Angel Investor is.

Special thanks to the GSV TEAM who are really like a family to me. They are bright, passionate, tireless and really nice. I had the idea for *The Global Silicon Valley Handbook* for a while but without Rahul Singireddy and Ryan Demo, this project would have gone nowhere. Rahul's amazing wit and writing made this book come alive, and Ryan's eye for design was exceptional.

To Suzee Han who has been my right-hand and who has helped me on every aspect of *The Global Silicon Valley Handbook* and literally got it over the goal-line. Among her many talents she can read my mind, which is both scary and efficient. Also thanks to Nick Franco—my Chief of Staff, Chief Researcher, Chief Advisor and Chief Critic—whose roles have been immeasurably helpful. And thanks to Li Jiang, GSVlab's self-proclaimed Chief Evangelist, who is the heart and soul of GSV and inspires me to never stop reaching for the stars.

Thanks to others at GSV who have especially put their shoulders to this effort including Matt Hanson, our man about town in the Big Apple; Luben Pampoulov, who does his scouting from Paris, but has a nose for ideas and people everywhere; Gopi Pepakayala, who is based in Bangalore, and leads our research efforts at GSViQ; GSV Venture's head Larry Aschebrook who in any given week could be in Dubai, Mumbai and Shanghai; Deborah Quazzo, who leads GSV Acceleration; and GSV Growth Credit's David Spreng.

I want to give a big thank you to Andy Goldenberg and Cliff Humphreys at BrandLift who have helped me and GSV in so many ways to really articulate what the Global Silicon Valley is about.

Another huge acknowledgement of appreciation to the GSV Family who have also contributed to *The Global Silicon Valley Handbook*: Debbie Elsen, Mark Flynn, Bill Tanona, Spencer McLeod, John Denniston, Mark Moe, Richard Harris, Michael Cohn, Diane Flynn, Sidarth Balaji, Michael Narea, Siya Raj Purohit, Maria Victoria Ferrara, Sandy Lin, Anthony Palma, Chidvi Pemula, Emily Ha, and Justine Hausner.

And thanks to the rest of the contributors who helped make the book a reality: Brian Abraham, Tom Adams, Tom Alexander, Jeanne Allen, Agnaldo Andrade, John Bailey, Anshumaan Bansal, Maxwell Barnes, Rodrigo Barros, Michael Bartimer, Peter Bell, Marc Benioff, Michael Boyers, Walter Buckley, Evan Burfield, Alex Capecelatro, Mike Carter, Steve Case, Lucas Ceschin, Sanjay Challa, Terence Chan, Anthony Chang, Audrey Cheng, Jessica Chitkuer, Grant Chou, James Citrin, John Coughlan, Michele D'Aliessi, Pieter de Jong, Perry Dellelce, Rahul Desai, Akankshu Dhawan, Pat Dillingham, Aidan Donohue, Tim Draper, Joe Ducar, Alex Ellis, Joel Eriksson, Chris Fernandes, Lisa Flynn, Patrick Franco, Nicole Franco, Jordan Frankfurt, Aaron Fu, Graham

Furlong, Amir Gelman, Sourish Ghosh, Garrett Goehring, Murray Goldberg, Jonnie Goodwin, Karen Greene, Alexandra Greenspan, Rusty Greiff, Sina Gritzuhn, Puja Gubbi, Vicky Guo, Justin Halsall, Albert Hanser, John Hartnett, Felipe Held, David Helfer, Andy Hidayat, Rand Hindi, Derek Hockman, David Hoffman, Drew Hoffman, Gordon Hoge, Marissa Hopkins Secreto, James Hu, Ricky Hunter, Lauri Järvilehto, Colleen Jiang, Victor Jiang, Jeremy Johnson, Tara Kelly, Karan Khemka, Charmaine Kng, Aneesha Kommineni, Dick Kramlich, Arun Kumar, Sunyoung Kye, May Lee, Jesse Levin, Mike Levinthal, Dan Levitan, Garros Li, Harry Li, Par Lindstrom, Marianne Lino, Joe Lonsdale, Ronnie Lott, Nancy Luo, Jeremy Lum, Christoffer Mailing, Pramath Malik, Andrew Marks, Anna Mason, Vaibhav Menon, Cindy Mi, Joey Mi, Nathan Millard, Jaime Montgomery, Ricardo Morales, Bryce Morisako, Mark Musolino, Karim Mustaghni, Jaime Nack, Margaret O'Mara, Evon Onusic, Dave Pottruck, Rudy Prince, Scott Prindle, Chris Protasweich, Tammer Qaddumi, Ben Quazzo, Louise Rogers, Audrey Russo, Jamal Saab, Bill Sahlman, Joe Sanberg, Mike Sanders, Dave Schiff, Joanna Schwartz, Lakshmi Shenoy, Ryan Shuken, Ben Slivka, Matt Sonsini, Matt Spector, Björn Stansvik, Lucy Stonehill, Kasper Suomalainen, Kelly Szejko, Jessica Tan, Stephen Tang, Aditya Tripathi, Howard Tullman, Tracy VanGrack, Anton Waitz, Ben Wallerstein, Patrick Walujo, Howard Wang, Todd Warren, Andy White, Joseph Wilkinson, Melissa Wooten, Kylie Wright-Ford, Dun Xiao, Sam Xiao, Kevin Zhou, William Zhou, Gabriella Zielke.

And finally, I want to thank the fantastic people at Grand Central who believed in the book, including my acquiring editor Gretchen Young, and to my editor Katherine Stopa. Thanks also to Jamie Raab, Beth de Guzman, Jeff Holt, Shelby Howick, and Tiffany Sanchez. And last, but certainly not least, thanks to Jillian Manus, who made this all happen.

A big thanks.

The Global SILICON VALLEY

GLOSSARY

A

A/B test

A strategy startups use to compare two versions of a product to see which one performs better: "Let's do an A/B test to see if changing the coffee to dark roast has an effect on productivity." For a more specific type of A/B test, see "Phantom testing."

Acquihire

Buying out a company for its talented workers as opposed to its products or services: "I don't want to be acquihired. I refuse to work for your company. This startup is my passion, dream, and...ten million dollars? Done."

Angel

An affluent individual that provides seed-stage funding: "She's not dead...She's just your guardian angel now, giving your startup funding from her will."

API

Acronym thrown around to appear techno. Stands for "application program interface" and specifies a set of routines, protocols, and tools for building software applications that interconnect with each other. The use of APIs allows you to order a Lyft for yourself or your colleagues through Slack.

AR/VR

Stands for "augmented reality/virtual reality". In AR, digital elements supplement a user's normal perception. In VR, users are fully immersed in a digital world. In other words, Pokémon Go is augmented reality because players run through real cities (and into real cars), whereas with virtual reality, you are just looking like an alien wearing goggles.

AU

Stands for "Active Users," most commonly used as MAU (Monthly Active Users) or DAU (Daily Active Users). Given the correlation between the frequency of engagement and valuation, founders are starting to use a new MAU, which stands for "minute active user," HAU, "hourly active user," and SAU, "second active user," as an up-and-coming metric as well.

AUM

Abbreviation for "Assets Under Management," the total market value of the financial assets a firm manages on behalf of their clients. "AUM is an investment firm's equivalent of comparing hand sizes, but the real players measure the return on the pile, not how big the pile is."

B

Bandwidth

Time and availability to do a project:

Boss: "Hey do you have the bandwidth to do this stupid, meaningless task that I don't want to do?"

Employee: "Of course I do, boss, I just won't sleep for a couple days."

Boss: "Great, for this kind of a work, you'll be a shoe-in for a promotion maybe."

BDR—Business Development Representative

New, more glamorous term for an old entry-level shit job: telesales. Enables Ivy League graduates who can't find a job to fool their parents and friends into thinking that the $200,000 they paid for their diploma wasn't flushed down the toilet.

Benchmark

Performance goals against which a company's success metrics are measured. Not to be confused with Valley titan Benchmark Capital. Commonly used by venture capitalists when deciding whether or not to invest in a company: "Since Valley startups rarely have revenue, let alone earnings, benchmarks can help investors think they are being rational when they should be scared out of their minds."

Big Data

A broad term for data analytics that by putting the adjective "big" in front implies massive insights. To date, Big Data has been closer to Big Foot, rumored to be a force but rarely seen. As a non-*Handbook* reader would say, "My Big Data app is the holy grail." (See 23)

Bleeding Edge

More advanced than "cutting edge" and bloodier for its victims as well: "Our bleeding edge virtual reality headset is cutting off our users heads."

Blockchain

On the surface, it is two bad words that together create a positive and highly disruptive idea. In actuality, it is a technology that allows for the transfer of information through a shared record or ledger of events, and removes the need for a third party to intermediate and validate the transfer. The most famous example of a blockchain is Bitcoin.

Bookings

The value that a consumer intends to pay the company for a service. This is usually done in a contract: "For example, a startup might have total bookings of $3 million with a customer over 3 years, so their bookings will be $3 million, but their actual revenue will be whatever they collected. It's a great way to make startups seem more real than they are, but bookings don't buy my kids shoes. Revenues do."

Bootstrap

Self-financing your startup with savings and credit cards as opposed to receiving investor funding: "I'm bootstrapping right now, but I should get some seed capital sooner or later. Living with my parents and working on my idea when not playing Call of Duty isn't too bad, though."

Bozo Explosion

Steve Jobs' lament that when early employees at startups get promoted over time to jobs they aren't qualified for and, additionally, hire people not as talented as themselves, you wake up one day and realize the company is run by a bunch of idiots. Bozo explosion often leads bozo implosion.

Brain Rape

When an investor or another company tries to mine your company secrets under the pretense of a pitch or other official meetup:

Engineer A: "Hey this is great, they love us! You should bring us to these meetings more often"

Engineer B: "Don't you get it, they're trying to steal our ideas and brain rape us."

Bridge

Raising capital typically between two formal funding rounds. Investors are cautioned to avoid a "bridge to nowhere," where the prospects of bringing in the next funding are dubious:

Person 1: "We're burning money way too fast—we can't sustain this to our next equity round. I think we'll need some bridge financing."

Person 2: "Wasn't our $250 million Series C yesterday?"

Brogrammer

Portmanteau of bro and programmer, a category of employees in the tech world who are typically young, male graduates with a "bro-y" culture:

Bro 1: "Let's get fucked up and code, dude. I'm talking three bottle pulls of margarita mix and Malibu."

Bro 2: "That's savage. You're a real brogrammer. Wish we drank in college."

Bubble

A state of booming economic activity (as in a stock market) not sustainable by underlying values that often ends in a sudden collapse: "While $7 trillion were destroyed in the 2000 Internet bubble, many in Silicon Valley pray every day for a new bubble."

Burn Rate

The amount of money a startup loses per month, which is typically compared to the amount of cash in the bank. In other words, if a company has $1.2 million in the bank and is "burning" $100,000 a month, you'd say they'd have twelve months of "burn" left. "Their burn rate increased to $300,000 a month after they hired their professional juggler for coffee breaks."

Burning Man

A weeklong event held in Black Rock Desert, Nevada, that promotes radical self-expression; the festival has an increasingly greater number of techies from Silicon Valley in attendance: "I used to go to Burning Man, but now it's just too corporate—there's VC's coming in from Silicon Valley in helicopters. I mean, I work at Facebook but at least I'm down to drop acid."

Business Attire

Means suit and tie in the business world, but has transformed to a more comfortable uniform of jeans and a T-shirt in the Valley (see "How to Dress," page 17): "I said business attire, team—jeans, not khakis."

BYOD

Bring your own device:

Person 1: "I heard I would get a ton of a perks if I worked in the Valley, but I don't even get a work computer."

Person 2: "Most startups are BYOD."

C

CAC—Customer Acquisition Cost

A metric used to determine the economic efficiency of bringing in a customer versus the cost of marketing and sales to attract the customers. Often used alongside Lifetime Value (LTV) of a customer: "Pets.com became Pets.bomb when people realized it cost a thousand dollars to bring in a paying customer versus the $30 the customer actually spent on a product."

Churn Rate

Percentage of subscribers discontinuing a subscription service. Rule of thumb, if the churn rate is above 30% per year, your business model is challenged: "Recurring revenue is the holy grail—ugh! Sorry (see "Phrases to Avoid in a Pitch," page 23), but if your customers drop your service like a bad habit and don't renew, trouble lies ahead." "High churn rate is only good if you're making butter."

Close
Signing all the legal papers and documents to finalize an investment. See Wikipedia for "It's not over until the fat lady sings."

Cloud
A network of remote servers hosted on the Internet and used to store, manage, and process data in place of local servers. "If an investor asks you why it works, just say it's in the cloud."

Critical Mass
The point at which a startup scaled to be adequate in size and became viewed as established and now has to worry about being disrupted by another startup: "Once Snapchat reached critical mass, there was a boom in various forms of disappearing communication apps."

Crowdfunding
Financing an endeavor with funding from many individual donors and investors through the Internet. Kickstarter and Angelist are examples. The wisdom in raising money from crowds is a double-edged sword. If things go great, you avoid the evil venture capitalist. If things don't, you now got a bunch of angry doctors, dentists, and lawyers as shareholders who know how to sue, inflict pain, and rip your teeth out: "I don't have to worry about venture capital money—I'll just crowdfund it."

Crowdsourcing
Utilizing the small contributions of many people for some end: "I wouldn't be the programmer I am today without the crowdsourcing available in websites like Stack Overflow. Turns out all my answers are from really bored inmates in the Cayman Islands."

Culture Tsar
A classic Silicon Valley title meant to be cool (see "The Silicon Valley Job Board", page 181), but unfortunately, anyone who knows history knows that most tsars were overthrown and killed in popular revolts.

D

Deliverable
Something tangible you owe for work: "What deliverables are you working on after this meeting? Let me add another."

Dilution
Percentage of ownership that is decreased when new investors come in: "Dilution should not be confused with delusional, although many founders have delusions about their dilution."

Disrupt
To address a new market or offer significant advantages to the current status quo of your class of products or services, leaving everyone in the existing market struggling to catch up: "This product is completely disrupting a trillion-dollar market with a billion users with early adopters like nothing that has ever come here before…ever."

Disruptive Technology
New technology that changes an industry's dynamic. A classic definition of a disruptive technology is if it provides 80% of the value at 20% the cost: "I miss my chauffer James, but Uber is a fraction of the cost and doesn't have to go home and see the fam."

Distorted Reality Field
A typical symptom of an entrepreneur who sees the world as what it can become, not what it is, but be-

lieves it already is that way: "The shrink says Suzee is crazy, but she's just operating with a distorted reality field."

Dogfooding

Using and relying on your own product, usually in beta version, to fix the kinks: "Brian Chesky is still dogfooding Airbnb—I heard he hasn't owned a home since 'Nam."

Down Round

Raising money at a lower valuation than a previous round, also thought to be worse than a terminal disease: "We're having a down round? Everyone's going to want equity like it's oxygen."

DRI—Directly Responsible Individual

Valley speak mumbo jumbo that basically says whose job it is to perform a task: "Chris is the DRI for cleaning the dishwasher."

Drinking the Kool-Aid

The process of sucking ordinary people into believing their company is truly extraordinary. Often associated with Silicon Valley leaders or a cult: "Jim has really drunk the Kool-Aid and believes that his game not only helps businesses but is saving humanity as well."

Due Diligence

Research and investigation into a company's figures, the period when an investor actually wants to see if the product or the people exist: "You can't complain that I misled you with my earnings report when you failed to do your due diligence."

E

EBITDA

Earnings before interest, tax, depreciation, and amortization, used as a general proxy to measure how much cash a company truly generates. Creative Valley companies try to novel financial metric of EBE (earnings before expenses): "It's not ebola. It's pronounced ee-bit-da."

Elevator Pitch

A short summary used to quickly and simply define the value proposition of a startup...In other words, you can make your company pitch in the time it takes to go from the lobby to the roof in an elevator. A fatal gaffe in your presentation is if you think it's a 100-story building, but in reality it's four floors. (see "Phrases to Avoid in a Pitch," page 23.)

Entrepreneur

A person who identifies a need and helps solve it by starting a business. Also see "Distorted Reality Field":

Person 1: "Hey, man, sorry you couldn't find a job this summer."

Person 2: "It's okay. I'll be an entrepreneur instead."

Escape Velocity

The point at which a startup is on the path for exponential growth: "Facebook seemed like it was all fun and games till it reached escape velocity in 2008. Now it's ubiquitous."

Evangelist

A person at a company who builds support for a product through fanatical and passionate belief. The job is not to do real work, but to wear a cool t-shirt and have a smile on their face: "Guy Kawasaki was the chief evangelist of Apple, popularizing its products through media, talks, and demonstrations."

Exit

When a startup is acquired or goes public. All exits aren't created equal (see MS-DOS):"Instagram was bought by Facebook in a one-billion-dollar exit."

F

Feng Shui (ph-eng sh-way)

The Chinese philosophical system of harmonizing everyone with their surrounding environment. The main reason why the executives at startups play the role of the interior decorator in an effort to create a more prosperous environment in the office. Also the reason why your desk randomly changes orientation on a weekly basis in an effort to attract stronger Chi.

Follow On

A subsequent investment made by an investor who had funded previous rounds. Founders note it's often a bad sign if your investors don't want to participate in a follow on financing. "Kleiner Perkins made a follow on investment after seeing the company's revenue skyrocket."

Founder Hounder

A person who pursues hookups with founders of start-ups. Hoodie Nation's version of a groupie.

Freemium

A product or service that is free for most consumers to use, but charges a fee for premium features. For example, air is free in Beijing, but a premium product to avoid dying is a mask.

G

Gig economy

The emerging business environment in which temp and freelance positions are common and organizations contract workers for short-term work: "Man, I love my new gig, and I'm going to love my next gig even more."

Glassholes

Originally a term to describe the dorky-looking nerds wearing the ill-fated Google Glass. Now used to describe any nerdy prick that thinks displaying any hipster-type technology is cool. "What a glasshole wearing his Fitbit, Apple Watch and Jawbone all on one wrist."

Growth Hackers

Marketers, engineers, and product managers that specifically focus on building and engaging the customers of the business to make it to the promised land: "To make it to our next financing, we need some real growth hackers."

Growth Metrics

Measurements and calculations that evaluate a startup's growth potential like MAU or MoM: "Looking at profit or current assets are death metrics—growth metrics are what really matter. Where is this company going to be?"

GSD—Get Shit Done

The most valued type of employee at most startups: "I could care less about his resume, Nick gets shit done." (see "Singapore," page 126.)

H

H-1B Visas

A non-immigrant visa, which allows U.S. companies

to temporarily employ workers in technical occupations: "Given the need in Silicon Valley for engineers, H-1B Visas are as valuable to a startup as a check from Sequoia."

Hacker

A person with an intellectual curiosity to build better things: "A hacker isn't someone who breaks into the confidential NSA network—that's what you call genius and not the ones in the Apple Store (see "The Silicon Valley Job Board", page 181),

Hockey Stick Growth

Weekly or monthly sales results that are vertical:

VC 1: "This startup is great—look at that beautiful hockey stick growth."

VC 2: "That was over one day."

I

Ideate

Portmanteau of idea and create, means to brainstorm and think: "I'm tired of class. Let's go ideate our trillion dollar market opportunity to cure cancer on Mars."

Incubator/Accelerator

A physical space or program that helps startups grow, usually providing them with mentorship, networking opportunities, and sometimes capital: "My parents yell at my sister to get a real job every day instead of working on her Uber for red Solo cups—you could call our house an incubator/accelerator."

Infobesity

So much information that it impedes progress: "I'm trying to overcome my infobesity with Weight Watchers; they limit me to reading ten tweets a day."

Internet of Things

A phenomenon where all devices are connected and communicate with each other, knowing what you want almost before you do: "In the morning my alarm clock signaled my pan to start frying bacon, which alerted my ventilation system to direct the bacon smell from my kitchen to my bedroom—welcome to the Internet of Things."

IPO

Initial Public Offering is the first sale of stock by a private company to the public. Historically, what the brass ring venture-backed private companies aspired to do. Today viewed as the monetization option of last resort: "We ran out of options, so my VCs are making me do an IPO."

J

J Curve

Common shape that materializes when plotting the value of a venture capital fund over time, dipping at first because of the failures of bad companies (lemons ripen faster) before turning upward when the good companies take off: "I kept hoping for the J curve to show up, but the value of my fund just went lower and lower and lower till it resembled an L curve."

JOBS Act

Stands for Jumpstart Our Business Startups Act, a law enacted to make it easier for emerging companies to get funding and go public. Too bad the JOBS Act had nothing to do with Steve Jobs, and IPOs continue to be few and far between.

K

KPI—Key Performance Indicators

More Silicon Valley mumbo jumbo that basically says

what a company needs to focus on to be successful or be toast: "The KPI for our CEO is to not run out of money."

L

Lead Investor

Person or firm that prices a deal and sets the terms: "You can always divorce your spouse, but getting rid of a lead investor is more difficult unless you're in New Jersey (see "The Worst Places for Innovation," page 121).

Lean Startup

Method for developing businesses focused on iterations of the product: "I made sure my company was on the caveman diet to achieve the lean startup."

M

M&A

Abbreviation for "Merger and Acquisition," which describes the process through which one company takes over or combines with another. "The AOL/Time Warner M&A was a disaster!"

MOFU—Middle of Funnel

Marketing term to determine leads that are prospects to be customers. See "TOFU." "Our Director of Marketing said our MOFU and MOFOs."

MoM

Month over month growth; could be revenues, could be users, could be over if it's not growing fast enough. "MOM! Where's my tie?"

MVP

Minimumly Viable Product. A product with just enough features to show early adopters its potential, providing a feedback loop to guide future improvements:

Product 1.0: "You're the MVP."

Product 1.1: "No, you're the real MVP."

N

NDA

Stands for nondisclosure agreement; legal contract that details confidential material not to be shared. In the real world, VCs view NDAs as being as useless as the lawyers who drafted them: "After signing an NDA at NASA, I got access to their most secret documents—transformers are real."

Negging

Treating a possible investor or employee badly to make you and your company seem wanted, confident, and lucrative.

Erlich: "See everyone wants us, but by shitting all over us, they try to bring our price down. But you shitting all over them counteracted them shitting all over us. You negged a neg."

Network Effects

Where adding a new member to a network adds exponential value: "The network effects of Tinder are crazy, where one new hot guy brings in 50 new gals."

Networking

The act of meeting others and pretending to be interested in what they do in hopes that they'll help you in the future. For tips, see ("Do's and Don'ts of Networking," page 24): "Look at how many LinkedIn connections I've made through networking!"

Nootropics

"Smart drugs" that improve cognitive function, memory, creativity or motivation, in healthy individuals.

The politically correct way to say "LSD".

O

Onboard

The process by which a new employee is introduced to and gets situated into a new company. "I can't do meetings today. I'm onboarding the new employees."

P

Peer-to-Peer

A computer network in which each device can act as a server for others, allowing shared access to files: "Brian Chesky is the Abraham of the peer-to-peer companies."

Phantom Testing

Often done when the company doesn't have a real product to test but wants to get customer leads and market information. "Phantom testing is a modern-day guinea pig experiment."

Pivot

When a startup changes direction but retains its team: "My startup pivoted from a coffee delivery app to a barista team at Starbucks."

Platform

A business that creates value by facilitating connections between two or more interdependent groups: "I'm not an app, I'm a platform business and I need to be treated as such."

Post-money

The value of a company after it receives its financing. Not to be confused with when the company has run out of money: "Joe's app post-money valuation was $13 million, after he received $3 million from angels."

Pre-IPO

Round of financing done right before a company goes public: "The real winners of most IPOs are private equity funds who get to buy stock at a pre-IPO price."

Pre-money

The value of the business before funding comes in. Not to be confused with before you're making any money: "The pre-money valuation for Joe's app was $10 million before he brought in $3 million of angel money, giving it a post-money valuation of $13 million."

Pre-revenue

The state of a company before it generates any revenue: "Valuing pre-revenue companies is an art—you basically take the number of employees and multiply it by a billion."

Preferences

Giving investors a priority on the return of invested capital, giving some guarantee of money back: "He has 1x preference, so if he invests $100 million on a $1 billion evaluation, and the company is later sold for $500 million, he is guaranteed his $100 million. If he has 2x preference, he'd be guaranteed $200 million."

Price to Sales Ratio

Dividing a company's market cap or valuation by its revenue: "My company's price to sales ratio is an imaginary number."

Profit Margins

Net income divided by net revenue, looking at how much the company has profited in relation to costs.

VC: "Can you explain how you're projecting your

profit margins?"

Entrepreneur: "We're actually using a more advanced accounting metric called EBITDASMBPER—earnings before interest, tax, depreciation, amortization, salary, marketing, benefits, perks, engineering, and research. In other words, earnings before expenses."

R

Rachet

Anti-dilution provision for investors typically in the later rounds to protect them from future financing: "This rachet is really rachet (pronounced *rat shit*)."

Recurring Revenue

The measure of sales that is ongoing in nature. This excludes the one-time fees associated with a product. Software-as-a-service company (SaaS), for example, gets paid monthly for each user, so their revenue is recurring until its customer stops using it. This is the opposite of recurring nightmare and in fact, allows entrepreneurs and investors to sleep better.

Retention

Keeping people associated with a company, whether in terms of users or employees: "Their retention rate is amazing—it's like the heroin of music subscription services."

ROI

Return on investment is the profit or loss percentage on the amount of capital invested: "Our ROI is huge!" If you're Mark Cuban a.k.a Russ Hanneman...ROI = Radio on Internet.

Runaway Valuation

A phenomenon that happens when multiple VCs need to be in the deal where the valuation can't be justified by the math. Risk gets defined by looking like a loser or missing out: "The over-interest of many VCs led to the company's runaway valuation."

S

Scale

The ability to expand to a meaningful size quickly: "Your company's doing great in the American two-year-old market, but can it scale to the Chinese two-year-old market? Do you have the infrastructure for that?"

Seed Round

The early capital used to help start a company; comes with a huge risk of company failure but also a huge reward if the company is successful: "Taking a huge bite of a watermelon, Peter Thiel spit out some black pits in a much publicized seed round."

Series A/B/C/D/E and later

Post-seed stage funding rounds that occur after companies develop their business past certain milestones. Companies are usually at a higher valuation each round, prompting investors to pay higher prices for later investment opportunities: "The startup is attempting to raise 26 individual Series rounds to rival Google; it was quoted as saying, 'You can do Alphabet, I can do Alphabet.'"

Sharing Economy

See "Peer-to-Peer," hybrid market model which refers to the peer-to-peer sharing of access to goods and services: "Millennials want to share everything and own nothing."

SMB

Stands for "small medium business," which is usually a company that has fewer than 100 employees. "The BDR's KPIs are on signing up more SMBs."

Stealth Mode

The state a startup is in when it is working secretly behind the scenes without drawing attention, usually to avoid competitors: "The company's in stealth mode right now, avoiding copycats, competitors, and paying customers—or it just failed."

SLA—Service Level Agreement

An agreement where a service is formally defined between the provider and the user, including aspects of scope, quality and responsibilities: "We do not have an SLA in place, so no more excuses for why we can't seem to shoot straight."

Steve Jobs

God:

"Steve Jobs,
Who art in heaven
Hallowed be thy name;
Thy iPhone come,
The competitors be done,
On earth as it is in Apple."

Sticky

Used to describe products that retain users: "Facebook is as sticky as Donald Trump's wig made of lion manes."

Stock Options

Benefits for employees in the form of options that can be exercised at a certain price; used to keep employees around longer: "I joined Box early, so my stock options were great. I got the ability to buy 10,000 shares at a price of $1.50—the price on the market for Box stock is now $100 three years later. Now I can buy 10,000 shares for $1.50 each even though they are now worth $100 each. I still have to worry about capital gains taxes, though."

Strategic

A company who is in the same industry as the startup they invested in. Strategics are alluring to startups, and they often pay a higher valuation and give credibility. Strategics utilize the naiveté of startups to steal their best ideas or squash them: "Because GE is in the energy industry, cleantech companies have to have them as a strategic investor."

Sunsetting

The extermination of a product or service: "Me and a couple buddies grabbed some brews and cigars and watched the sad, beautiful sunsetting of Google+. It's not you Google+. It's me."

Syndicate

A loose coalition of investors working together to handle some large deal or transaction: "Google Ventures and Benchmark teamed up as the investment syndicate for an up-and-coming startup."

T

Takeunder

An offer to purchase a company for less than its market price. This is Option Z in the entrepreneur's playbook: "After receiving a ton of lawsuits, the company accepted a takeunder."

TAM

Total Addressable Market. Refers to the amount of revenue that could be available for the product based on the markets it could impact: "The founders were great, but the TAM was too small for the product—there's only so many nuns who vape in the world."

-tech

An abbreviation for the industry a startup is in or entrepreneur is involved with. Most common ones include Biotech, Edtech, Fintech, but terms like Agtech are also on the rise. Given the ubiquitous reality of tech, putting tech at the end of an industry is a bit redundant. Kind of like an "unexpected surprise."

Person 1: "What are you interested in?"
Person 2: "Edtech is my true passion!"

TED Talks

The cult millennial conference where the themes of Technology, Entertainment, and Design converge in the form of short, powerful talks. Some of Silicon Valley's best have taken the TED stage. Did you watch the TED Talk on "10 things you didn't know about orgasms?"

TOFU—Top of Funnel

Marketing term to identify all the potential leads for a product or service (see MOFU): "We have a TOFU problem because the number of leads is smaller than the number of our employees."

Tourist

Mutual funds, hedge funds, or corporations that do not specialize in venture capital but occasionally make a private market investment: "When I saw that private equity firm with a Hawaiian shirt, fanny pack, and $300 million dollar bet on GrabTaxi, I knew it was a tourist."

Tweetstorm

A series of posts on Twitter that are rapidly posted back to back for at least 10 posts. Notable Tweeters in Silicon Valley famous for their tweetstorms include Marc Andreessen and Bill Gurley, also known as Silicon Valley's Seinfeld and Newman.

Person 1: "Did you see Marc Andreessen's tweetstorm today?"
Person 2: "Which one? The one he posted at 3:10 p.m. or 3:14 p.m.?"

U

Ubercorn

A private company worth over $10 billion, with the name inspired by Uber's huge valuation:

Person 1: "Horses were disrupted by cars—now it's time for horses to strike back. Welcome to the Ubercorn."
Person 2: "Isn't that just Travis Kalanick on a horse?"

UI

Stands for User Interface, or the design of a product's consumer facing platform: "Craiglist's UI has to be one of the worst I've seen on the Internet." Pundits love to criticize a startup's UI, but the market will decide. Don't believe us? Look at Spotify.

UX

Stands for User Experience, or the design of the experience consumers have while on a product's platform: "Airbnb's UX and UI are top notch."

Unicorn

A private company worth over $1 billion: "The Unicorn is a rare type of horse with a horn, most often seen in the dreams of girls under six."

Unicorpse

A Unicorn that goes out of business: "With all these

Unicorns popping up, some of them are fated to be unicorpses."

Uniscorn

A Unicorn that was once dubbed the "next hottest thing" that blows up in a way that makes its investors embarrassed to think about it. "Don't ever mention Webvan, the original Uniscorn on Sand Hill Road."

Unit Cost

Cost incurred by a company for one unit of its product: "The unit cost was supposed to decrease with time, but kept going higher because of confounding variables—yacht parties."

V

Value Add

Describes additional benefits the company gives to increase the product's price or value: "My new iPhone came with the value-added feature of my imaginary lover: Siri."

Venture Capitalist

An investor that takes higher risk to fund startups. Young VCs try to look older than they are...Old VCs do exactly the opposite. Both still wear khakis.

Entrepreneur: "Look, there's a venture capitalist... Does he write checks?"

Venture Debt

An oxymoron or potentially smart way to get less dilutive capital before hitting milestones: "To reach the milestone of a billion users, the company took on venture debt from City National Bank."

Viral

Growth spreading exponentially and often globally—a powerful characteristic of network effects. Often mistaken as a sexually transmitted disease, but actually something founders want.

W

Wantrapreneur

A person who talks a big game about being the next Mark Zuckerberg and building the next Facebook but spends all of his time reading TechCrunch and watching *Shark Tank*.

Warrants

The right to purchase securities at a specific price within a certain timeframe. Founders who steal money or make a fraudulent promise deal with a different type of warrant: "The company enticed investors with more warrants, allowing them to buy more of the company for less."

Waterfall

A sequential flow of money. While waterfalls project this image of blissful tranquility, if you're at the bottom of a waterfall, you might as well jump off the top of one: "In a private equity fund, the investors usually get back the money they invested because that's at the top of the distribution waterfall."

Work-life Balance

Idea of work and life being separate entities, with some time devoted to each: Everyone in the Valley: "What's work-life balance?"

Bibliography

"10 Facts about Hamburg." Globalblue. Accessed December 13, 2016. http://www.globalblue.com/destinations/germany/hamburg/10-facts-about-hamburg.

"10 Facts About Hong Kong." USA Today. Accessed December 13, 2016. http://traveltips.usatoday.com/10-hong-kong-21270.html.

"10 Facts about Tokyo." Globalblue. Accessed December 13, 2016. http://www.globalblue.com/destinations/japan/tokyo/10-facts-about-tokyo/#slide-4.

"10 Interesting Facts about Dubai." The Crazy Facts. August 23, 2014. Accessed December 13, 2016. http://www.thecrazyfacts.com/10-interesting-facts-dubai/2/.

"10 Interesting Facts About Hamburg." OhFact! December 02, 2016. Accessed December 13, 2016. http://ohfact.com/interesting-facts-about-hamburg/.

"10 Interesting Facts About Hong Kong." UNIGLOBE Direct Travel Ltd. Accessed December 13, 2016. http://www.uniglobedirect.com/view-vacation.html?id=10-interesting-facts-about-hong-kong.

"10 Unbelievable And Cool Facts About Detroit." Daily Detroit. June 12, 2016. Accessed December 13, 2016. http://www.dailydetroit.com/2014/12/01/10-unbelievable-cool-facts-detroit/.

"11 Cool Facts You Didn't Know About Dublin, Ireland." EscapeHere. Accessed December 13, 2016. http://www.escapehere.com/destination/11-cool-facts-you-didnt-know-about-dublin-ireland/.

"11 Facts About Nashville You Never Knew Were True." OnlyInYourState. Accessed December 13, 2016. http://www.onlyinyourstate.com/tennessee/nashville/nashville-facts/.

"11 Fun Facts about Nashville." A Southern Gypsy's Adventures. April 26, 2016. Accessed December 13, 2016. http://asoutherngypsy.com/11-fun-facts-about-nashville/.

"12 Fun Facts About Kuala Lumpur, Malaysia." Traveling with the Jones. Accessed December 13, 2016. http://www.travelingwiththejones.com/2014/12/10/12-fun-facts-about-kuala-lumpur-malaysia/.

"15 Interesting Facts About Jakarta." OhFact! July 05, 2016. Accessed December 13, 2016. http://ohfact.com/interesting-facts-about-jakarta/.

"15 Interesting Facts About Kuala Lumpur." OhFact! July 25, 2016. Accessed December 13, 2016. http://ohfact.com/interesting-facts-about-kuala-lumpur/.

"18 Facts About Pittsburgh You Never Knew Were True." OnlyInYourState. Accessed December 13, 2016. http://www.onlyinyourstate.com/pennsylvania/pittsburgh/fun-facts-pittsburgh/.

"20 Interesting Facts About Copenhagen, Denmark." Travel Inspiration 360. November 30, 2016. Accessed December 13, 2016. http://www.travelinspiration360.com/20-interesting-facts-about-copenhagen-denmark/.

"20 Reasons Lagos Is An Amazing Place To Live." Travelstart Nigeria's Travel Blog. October 03, 2014. Accessed December 13, 2016. http://www.travelstart.com.ng/blog/20-reasons-lagos-amazing-place-live/.

"20 Things You Didn't Know about Salt Lake City." Matador Network. Accessed December 13, 2016. http://matadornetwork.com/trips/20-things-you-didnt-know-about-salt-lake-city/.

"20 Things You Didn't Know about Salt Lake City." Matador Network. Accessed December 13, 2016. http://matadornetwork.com/trips/20-things-you-didnt-know-about-salt-lake-city/.

2016 Wilmer Hale IPO Report. Report. March 24, 2016. Accessed November 21, 2016. http://www.wilmerhale.com/uploadedFiles/Shared_Content/Editorial/Publications/Documents/2016-WilmerHale-IPO-Report.pdf.

"21 Facts About Vancouver You Probably Didn't Know." VGC International College. October 11, 2016. Accessed December 13, 2016. http://www.vgc.ca/21-facts-vancouver-probably-didnt-know/.

"22 Interesting Facts About Dubai That Are Larger Than Life!" Urban Cocktail. February 10, 2016. Accessed December 13, 2016. https://grabhouse.com/urbancocktail/interesting-facts-about-dubai-that-are-larger-than-life/.

"25 Things You Didn't Know About Philadelphia." Mental Floss. Accessed December 13, 2016. http://mentalfloss.com/article/55036/25-things-you-didnt-know-about-philadelphia.

"25 Things You Should Know About Copenhagen." Mental Floss. Accessed December 13, 2016. http://mentalfloss.com/article/74685/25-things-you-should-know-about-copenhagen.

"33 Things You Probably Didn't Know About Detroit." Movoto. http://www.movoto.com/guide/detroit-mi/detroit-facts/.

"37 Things You Probably Didn't Know About Salt Lake City." Movoto. Accessed December 13, 2016. http://www.movoto.com/guide/salt-lake-city-ut/salt-lake-city-facts/.

"50 Fun Facts about Moscow." Friendly Local Guides Blog. June 15, 2016. Accessed December 13, 2016. https://friendlylocalguides.com/blog/fun-facts-50-facts-about-moscow.

"50 Interesting Facts About Sydney." Sydney. Accessed December 13, 2016. http://www.weekendnotes.com/interesting-facts-about-sydney/.

"55 Interesting Facts about Hong Kong." Weekend Notes. http://www.weekendnotes.com/interesting-facts-hong-kong/.

Austin, Scott, Chris Canipe, and Sarah Slobin. "The Billion Dollar Startup Club." WSJ. Accessed November 21, 2016. http://graphics.wsj.com/billion-dollar-club/.

"The Best Colleges in America, Ranked." College Rankings and Data | US News Education. Accessed November 21, 2016. http://colleges.usnews.rankingsandreviews.com/best-colleges.

"Billie Jean King Quotes." BrainyQuote. Accessed December 14, 2016. http://www.brainyquote.com/quotes/quotes/b/billiejean121917.html.

Cogman, David, and Alan Lau. "The 'tech Bubble' Puzzle." McKinsey & Company. Accessed November 21, 2016. http://www.mckinsey.com/business-functions/strategy-and-corporate-finance/our-insights/the-tech-bubble-puzzle.

Commentary. "South Africa's Startups Are Turning out to Be the Silver Lining in a Struggling Economy." Quartz. January 25, 2016. Accessed December 14, 2016. http://qz.com/601736/theres-a-silver-lining-in-south-africas-startup-scene-even-as-the-economy-struggles/.

CrunchBase. "The CrunchBase Unicorn Leaderboard." TechCrunch. Accessed December 06, 2016. https://techcrunch.com/unicorn-leaderboard/.

Deloitte Research. *Rethinking the Role of IT for CPG Companies.* Report. Accessed November 21, 2016. https://www2.deloitte.com/content/dam/Deloitte/us/Documents/consumer-business/us-cp-rethinking-the-roleof-it-042512.pdf.

"Did You Know? Interesting Facts about Helsinki, Finland! - Tallinn Tours &." Tallinn Tours & Tallinn Shore Excursions. November 25, 2015. Accessed December 13, 2016. http://estonianexperience.com/blog/2015/11/20/know-interesting-facts-helsinki-finland/.

Ericsson Mobility Report. Report. Accessed November 21, 2016. http://www.ericsson.com/res/docs/2015/ericsson-mobility-report-june-2015.pdf.

Forbes. Accessed December 14, 2016. http://www.forbes.com/sites/bryanstolle/2014/07/22/vision-without-execution-is-just-hallucination/.

Frankfurt, Sprachcaffe. "GEOS Languages Plus - Vancouver." 10 Cool Facts About Vancouver. Accessed December 13, 2016. http://www.geosvancouver.com/visiting-vancouver/10 cool facts about vancouver.htm.

Frankfurt, Sprachcaffe. "GEOS Languages Plus, Montreal." 10 Facts About Montreal | GEOS Montreal. Accessed December 13, 2016. http://www.geosmontreal.com/visiting-montreal/10-facts-about-montreal.htm.

"Fun Facts You Might Not Know About Montréal!" Fun Facts You Might Not Know About Montreal. Accessed December 13, 2016. http://www.ishlt.org/ContentDocuments/2013AprLinks_MontrealFunFacts.html.

"Hamburg Startups, Led by FinTech, Catch up to Berlin Tech." Geektime. March 17, 2016. Accessed December 14, 2016. http://www.geektime.com/2016/03/17/fintech-dominates-among-hamburg-startups-as-the-city-catches-up-to-berlin-tech/.

"How to Start a Startup." How to Start a Startup. Accessed November 21, 2016. http://startupclass.samaltman.com/.

"Interesting Facts About Copenhagen, Denmark." USA Today. Accessed December 13, 2016. http://traveltips.usatoday.com/interesting-copenhagen-denmark-21406.html.

"Interesting Facts about Portland Oregon - AfterGlob." AfterGlobe. December 23, 2013. Accessed December 13, 2016. http://afterglobe.net/interesting-facts-about-portland-oregon/.

"Investors Have Learned from Nasdaq 5K." CNNMoney. Accessed November 21, 2016. http://money.cnn.com/2004/03/09/technology/techinvestor/lamonica/.

"Israel Startup Map." Mapped In Israel. Accessed November 21, 2016. https://mappedinisrael.com/.

"Jack Ma Quotes." BrainyQuote. Accessed December 14, 2016. http://www.brainyquote.com/quotes/quotes/j/jackma678619.html.

Kauffmann, Laura. "10 Astounding Facts You Didn't Know About Madrid." Culture Trip. Accessed December 13, 2016. https://theculturetrip.com/europe/spain/articles/10-astounding-facts-you-didnt-know-about-madrid/.

Kharas, Homi. "The Emerging Middle Class in Developing Countries." OECD Development Centre, January 2010. Accessed November 21, 2016. https://www.oecd.org/dev/44457738.pdf.

"Martin Luther King, Jr. Quotes." BrainyQuote. Accessed December 14, 2016. http://www.brainyquote.com/quotes/quotes/m/martinluth106169.html.

McGee, Jamie. "Startup Dilemma: To Stay or Leave Nashville." The Tennessean. November 27, 2016. Accessed December 14, 2016. http://www.tennessean.com/story/money/2016/11/27/startup-dilemma-to-stay-or-leave-nashville/93450714/.

Meeker, Mary. "2016 Internet Trends Report." Kleiner Perkins Caufield Byers. Accessed November 21, 2016. http://www.kpcb.com/blog/2016-internet-trends-report.

Meena, Satish. "Consumers Will Download More Than 226 Billion Apps In 2015 | Forrester Blogs." Forrester Research. Accessed November 21, 2016. http://blogs.forrester.com/satish_meena/15-06-22-consumers_will_download_more_than_226_billion_apps_in_2015.

"Michelangelo Buonarroti Quotes." Goodreads. Accessed December 14, 2016. http://www.goodreads.com/author/quotes/182763.Michelangelo_Buonarroti.

National Venture Capital Association. *National Venture Capital Association.* Report. March 2016. Accessed November 21, 2016. http://nvca.org/research/stats-studies/.

PitchBook Data. *Pitchbook Universities Report.* Report. September 6, 2016. Accessed November 21, 2016. http://files.pitchbook.com/pdf/PitchBook_Universities_Report_2016-2017_Edition.pdf.

"A Quote by Abraham Lincoln." Goodreads. Accessed December 14, 2016. http://www.goodreads.com/quotes/328848-the-best-way-to-predict-your-future-is-to-create.

"A Quote by Steve Jobs." Goodreads. Accessed December 14, 2016. http://www.goodreads.com/quotes/513460-why-join-the-navy-if-you-can-be-a-pirate.

"Quotes." John Wayne. Accessed December 14, 2016. http://johnwayne.com/legacy/quotes/.

Schwaff, Angelika. "Fun Facts Hamburg." - MEININGER Hotels. December 11, 2014. Accessed December 13, 2016. http://www.meininger-hotels.com/blog/en/fun-facts-hamburg/.

"Screw It, Let's Do It." Virgin. December 12, 2015. Accessed December 14, 2016. https://www.virgin.com/richard-branson/books/screw-it-lets-do-it.

S&P Global Market Intelligence. Accessed November 21, 2016. www.capitaliq.com.

"Startupquote." Startup Quote. November 20, 2010. Accessed December 14, 2016. http://startupquote.com/post/1624569753.

"Startups Solving Indonesia's Big Problems Are Next $1b Company: Investors | Jakarta Globe." Jakarta Globe. Accessed December 14, 2016. http://jakartaglobe.id/tech/startups-solving-indonesias-big-problems-next-1b-company-investors/.

"Sydney to Launch Precinct for Thousands of Entrepreneurs as NSW Commits $13 Million to Startups - StartupSmart." StartupSmart. December 02, 2016. Accessed December 14, 2016. http://www.startupsmart.com.au/news-analysis/sydney-to-launch-precinct-for-thousands-of-entrepreneurs-as-nsw-commits-13-million-to-startups/.

Teare, Gené, and Ned Desmond. "Female Founders On An Upward Trend, According To CrunchBase." TechCrunch. May 26, 2015. Accessed November 21, 2016. https://techcrunch.com/2015/05/26/female-founders-on-an-upward-trend-according-to-crunchbase/.

"Ten Fun Facts about Sydney, Australia." UDrive Car Hire Blog. July 01, 2014. Accessed December 13, 2016. http://www.udrive.com.au/blog/facts-about-sydney-

australia/.
"Tokyo - 9 Facts About the Most Fascinating and Bizarre City in the World." Just One Way Ticket. February 19, 2013. Accessed December 13, 2016. http://www.justonewayticket.com/2013/02/19/tokyo-9-facts/.
"Top 10 Unbelievable Facts About Dubai." TheRichest. Accessed December 13, 2016. http://www.therichest.com/rich-list/the-biggest/top-10-unbelievable-facts-about-dubai/.
"Top 100 Inspirational Quotes." Forbes. Accessed December 14, 2016. http://www.forbes.com/sites/kevinkruse/2013/05/28/inspirational-quotes/#783f8c7d6697.
"Tweetwars: The Social Challenge in Twitter 'capital', Indonesia." Reuters. February 12, 2016. Accessed December 13, 2016. http://www.reuters.com/article/us-twitter-indonesia-idUSKCN0VK2LP.
"Venture Capital Database." CB Insights. Accessed November 21, 2016. https://www.cbinsights.com/.
Weinstein, Louis. "Bay Area to Standard American English Translator." McSweeney's Internet Tendency. Accessed November 21, 2016. https://www.mcsweeneys.net/articles/bay-area-to-standard-american-english-translator.
"Why Vancouver? Start-up Tips from Grow Conference." Hootsuite. Accessed December 14, 2016. http://www.slideshare.net/hootsuite/why-vancouver-startup-tips-from-grow-conference.
"Yahoo Finance - Business Finance, Stock Market, Quotes, News." Yahoo! Accessed December 14, 2016. https://finance.yahoo.com/.

Photo Credits

Page 2, Dallas City skyline at sunset, Texas, USA, dibrova/Shutterstock.com; Waterfront views of Tel Aviv, Dance60/Shutterstock.com; Tokyo cityscape and Mountain Fuji in Japan, jiratto/Shutterstock.com; New York City skyline with urban skyscrapers at sunset, USA, ventdusud/Shutterstock.com; MUMBAI, INDIA—17 JANUARY 2015: Chhatrapati Shivaji Terminus at sunset. It serves as headquarters of the Central Railways, Paul Prescott/Shutterstock.com. Pg. 3, Traditional old buildings in Amsterdam, the Netherlands, S. Borisov/Shutterstock.com; Dusk Chinese ancient buildings under the sky background (Nanchang Poetic), gyn9037/Shutterstock.com; Downtown Atlanta, Georgia, USA skyline, ESB Professional/Shutterstock.com; London at night with urban architectures and Tower Bridge, ESB Professional/Shutterstock.com; Nashville, Tennessee, USA downtown skyline on the Cumberland River, ESB Professional/Shutterstock.com. Pg. 4, Pioneer Hotel, Ryan Demo/GSV Asset Management. Pg. 5, Michael Moe Headshot, Ryan Demo/GSV Asset Management. Pg. 11, GSV 4P Star, GSV Asset Management. Pg. 12, Silicon Globe, GSV Asset Management. Pg. 16, An expansive panorama of the South San Francisco Bay area including the downtown San Jose skyline from Coyote Peak, Jeffrey T. Kreulen/Shutterstock.com; Pg. 17, The Founder, Ryan Demo/GSV Asset Management; The Investment Banker, Suzee Han/GSV Asset Management. Pg. 18, The Engineer, Suzee Han/GSV Asset Management; The Lawyer, Nick Franco/GSV Asset Management. Pg. 19, The Venture Capitalist, Suzee Han/GSV Asset Management; Pair of road cycling shoes on white background, badins/Shutterstock.com; Blue male leather loafers pair isolated on white background, Yevgen Romanenko/Shutterstock.com; Mens summer brown leather sandals isolated on white background, andregric/Shutterstock.com, Athletic shoes, Albanili/Shutterstock.com, Male person with hairy legs, walking towards, against white background, Arve Bettum/Shutterstock.com. Pg. 22, STANFORD, CA/USA—NOVEMBER 11: view of historic Standford University campus seen on November 11, 2013, California, USA, turtix/Shutterstock.com; Stanford Logo, Stanford University. Pg. 24, The Old Pro, Ryan Demo/GSV Asset Management. Pg. 25, Bucks in Woodside, Li Jiang/GSV Asset Management; Rosewood Sandhill, Li Jiang/GSV Asset Management; ; Village Pub, Suzee Han/GSV Asset ManagementPhilz, Ryan Demo/GSV Asset Management. Pg. 26, CHICAGO—FEBRUARY 8: A Toyota Prius C on display at the Chicago Auto Show media preview February 8, 2013 in Chicago, Illinois, Darren Brode/Shutterstock.com; Hong Kong, China Feb 6, 2013: Tesla Model S Electronic Car test drive on Feb 6 2013 in Hong Kong, Teddy Leung/Shutterstock.com; Contemporary Car Elegance Vehicle Transportation Luxury Performance Concept, Contemporary Car Elegance Vehicle Transportation Luxury Performance Concept/Shutterstock.com; Metal model car with a sun roof, HSNphotography/Shutterstock.com; ISTANBUL, TURKEY—MAY 21, 2015: Land Rover Range Rover in Istanbul Autoshow 2015, EvrenKalinbacak/Shutterstock.com; SANTAGATA BOLOGNESE, BOLOGNA, ITALY—JAN 20—Toy Lamborghini hurricane on white background, Tuesday 20 January 2015, Hetman Bohdan/Shutterstock.com; Horse and nice vintage coach with big yellow wheels. Isolated on white, cosma/Shutterstock.com; black sport bike on a white background, MiloVad/Shutterstock.com; DETROIT—JANUARY 13: The BMW M2 Coupe on display at the North American International Auto Show media preview January 11, 2016 in Detroit, Michigan, Darren Brode/Shutterstock.com. Pg. 29, Google Maps. Pg. 31, View from Alamo Square at twilight, San Francisco, prochasson frederic/Shutterstock.com. Pg. 33, Judgmental Map of Silicon Valley, GSV Asset Management; Judgmental Map of San Francisco, GSV Asset Management. Pg. 34, San Francisco Cable Car on California Street, November 13 2015, San Francisco USA, Pung/Shutterstock.com; Island and lake, Ashton Burton/Shutterstock.com; Mason Street houses, San Francisco, Gary Denham/flickr.com. Pg. 35, GSVlabs Building, Krista Paolella/GSVlabs. Pg.36, 2015 Pioneer Summit Team, Krista Paolella/GSVlabs; Joe Lonsdale at 2017 Pioneer Summit, Krista Paolella/GSVlabs. Pg. 37, Bill Campbell Power Pose, GSVlabs. Pg. 38, Office Slide Edited, Ryan Demo/GSV Asset Management; Donut Table, Li Jiang/GSV Asset Management; Trailer Home in Office, Li Jiang/GSV Asset Management. Pg. 39, Ping Pong Chandelier, Li Jiang/GSV Asset Management; Chess and Fooseball, Chegg; Office Bar, Li Jiang/GSV Asset Management; Coursera Wall, Coursera; Chegg Chair, Chegg; Flexport Cubbies, Li Jiang/GSV Asset Management; Couches, DogVacay; Putt Green, Chegg; Bunk Bed, Chegg; Bear and Bag, Chegg; Oval Office at Github, Li Jiang/GSV Asset Management. Pg. 40, Top Accelerator Map, Suzee Han/GSV Asset Management. Pg. 41, Spotify Logo, Spotify. Pg. 42, Facebook Logo, Facebook. Pg. 43, Twitter Logo, Twitter. Pg. 44, New York City skyline with urban skyscrapers at sunset, USA, ventdusud/Shutterstock.com. Pg. 47, Portrait of senior business man with ostrich egg, Rayman/Gettyimages.com; Hot Dog Truck, NYC, Tony Fisher/Flickr.com; Istanbul, Turkey - 15/06/2014: street musicians are present in a lot of streets of Istanbul and tourists are accompanied by their music during the visit of the city, COLOMBO NICOLA/Shutterstock.com; SEGOVIA, SPAIN - DECEMBER 12TH 2015: group of Asian tourists taking photographs, Kike Fernandez/Shutterstock.com; Attractive female freelancer hold smart phone while sitting at wooden table front open computer in modern coffee shop,young creative woman work on laptop while having breakfast on terrace, flare sun, GaudiLab/Shutterstock.com; Taxi Driver, Jim Pennucci/Flickr.com. Pg. 49, Downtown Los Angeles, California, ESB Professional/Shutterstock.com; Pg. 50, Snapchat Logo, Snapchat; Pg. 51, Tinder Profile, GSV Asset Management; Tinder Picture, GSV Asset Management. Pg. 53, Facebook Logo, Facebook; Groupon Logo, Groupon; Yahoo Logo, Yahoo; Twitter Logo, Twitter. Pg. 54, Snapchat Guide, GSV Asset Management; Snapchat Shot, GSV Asset Management. Pg. 55, Downtown San Diego at night, S.Borisov/Shutterstock.com. Pg. 59, Golden Gate at dawn surrounded by fog, Francesco Carucci/Shutterstock.com. Pg. 61, PALO ALTO, CA 9 NOVEMBER 2013 Considered the birthplace of Silicon Valley, the garage where Hewlett-Packard was founded is now a museum listed on the National Register of Historic Places, EQRoy/ Shutterstock.com; Google bicycle, akoppo/Shutterstock.com; Satellite dish pointed to the sky, sp.VVK/Shutterstock.com; MENLO PARK, CA—MARCH 18: A sign at the entrance to the Facebook World Headquarters located in Menlo Park, California on March 18, 2014. Facebook is a popular online social networking service, Katherine Welles/Shutterstock.com; LOS ALTOS CA / USA NOVEMBER 12: Famous house with garage where Steve Jobs and Steve Wozniak assembled their first computer the Apple I in 1976, turtix/Shutterstock.com. Pg. 64, London at night with urban architectures and Tower Bridge, ESB Professional/Shutterstock.com. Pg. 66, LONDON—AUG 3: Interior of pub, for drinking and socializing, focal point of the community, on Aug 3, 2010, London, UK. Pub business, now about 53,500 pubs in the UK, has been declining every year, Bikeworldtravel/Shutterstock.com; London city skyline silhouette background, Ray of Light/Shutterstock.com. Pg. 73, Dusk Chinese ancient buildings under the sky background (Nanchang Poetic), gyn9037/Shutterstock.com; Shanghai skyline with historical Waibaidu bridge, China, dibrova/Shutterstock.com. Pg. 79, Jack Ma Headshot, Alibaba Group. Pg. 82, China Map, GSV Asset Management. Pg. 84, Bill Campbell, GSV Asset Management; Steve Jobs, WikiCommons; Donna Dubinsky, GSVlabs; Marc Benioff, Salesforce; Ron Johnson, GSVlabs; Tim Cook, Apple. Pg. 85, Bill Campbell, GSVlabs; Justin Kitch, GSVlabs, Marc Andreessen, Andreessen Horowitz, Eric Schmidt, Alphabet, Jeff Bezos, Amazon, Danny Shader, GSVlabs, Mike McCue, ASU GSV Summit, Dan Rosensweig, ASU GSV Summit; Ben Horowitz, Andreessen Horowitz; Larry Page, Alphabet. Pg. 86, Chicago skyline. Chicago downtown skyline at dusk, Pigprox/ Shutterstock.com. Pg. 88, PayPal logo, PayPal; University of Illinois Logo, University of Illinois. Pg. 89, Lollapalooza Logo, Lollapalooza; EPIC logo, EPIC. Pg. 90, Berlin skyline panorama during sunset, Germany, Mapics/Shutterstock.com. Pg. 92, Rocket Internet logo, Rocket Internet; SoundCloud logo, SoundCloud; U8 logo, U8. Pg. 93, Waterfront views of Tel Aviv, Dance60/Shutterstock.com. Pg. 96, Israel Map, GSV Asset Management. Pg. 100, View on Eiffel tower at sunset, Paris, France, S. Borisov/Shutterstock.com. Pg. 102, Paris 11 April 2015: facade of the cozy cafes in Paris, yari2000/Shutterstock.com. Pg. 103, Europe Map, GSV Asset Management. Pg. 104, Toronto skyline at night in Ontario, Canada, Javen/Shutterstock.com. Pg. 106, Hack the North Logo, Hack the North. Pg. 108, view of Austin, Texas downtown skyline, f11photo/Shutterstock.com. Pg. 110, AUSTIN, TEXAS—FEBRUARY 3 2014: Bars, restaurants and other businesses in the Sixth Street Historic District, a major tourist destination that is listed in the National Register of Historic Places, Stephen B. Goodwin/Shutterstock.com. Pg. 112, BANGALORE INDIA—MARCH 3: Bagmane Tech Park is a software technology park equipped with all modern class facilities and is surrounded by a pond near the entrance. On March 3, 2011 Bangalore, India, Noppasin/Shutterstock.com. Pg. 115, India Map, GSV Asset Management. Pg. 116, Sundar Pichai Headshot, Alphabet; Indra Nooyi Headshot, Pepsi; Shannon Narayan Headshot, Adobe; Satya Nadella, Microsoft. Pg. 117, Most famous bridge in the city of Sao Paulo, Brazil, Celso Diniz/ Shutterstock.com. Pg. 120, CARNIVAL RIO DE JANEIRO—FEBRUARY 19: Samba School parade float at the Sambadome February 19, 2012, in Rio de Janeiro, Brazil. The Rio Carnival is the biggest carnival in the world, Leanne Vorrias/Shutterstock.com. Pg. 121, explosion nuclear bomb in ocean, Romolo Tavani/Shutterstock.com. Pg. 124, Seattle Skyline, TomKli/Shutterstock.com. Pg. 126, Singapore city skyline, Singapore Marina Bay cityscape at night, Noppasin/Shutterstock.com. Pg. 128, Traditional old buildings in Amsterdam, the Netherlands, S. Borisov/Shutterstock.com. Pg. 130, Denver Skyline with City Park in Foreground at Sunset, Albert Pego/ Shutterstock.com. Pg. 132, The White House on a beautiful summer day, Washington, DC, ESB Professional/Shutterstock.com. Pg. 134, Downtown of Miami, Florida, USA, ESB Professional/Shutterstock.com. Pg. 136, MUMBAI, INDIA—17 JANUARY 2015: Chhatrapati Shivaji Terminus at sunset. It serves as headquarters of the Central Railways, Paul Prescott/Shutterstock.com. Pg. 138, Downtown Atlanta, Georgia, USA skyline, ESB Professional/Shutterstock.com. Pg. 140, Bongeunsa Temple in the Gangnam District of Seoul, Korea, ESB Professional/Shutterstock.com. Pg. 142, Gamla Stan (old town) at night in Stockholm, Sweden, Mapics/Shutterstock.com. Pg. 145, Vinod Khosla, ASU GSV Summit. Pg. 147, Ron Conway, GSV Asset Management. Pg. 148, Humayun's Tomb, New Delhi, India, saiko3p/Shutterstock.com. Pg. 150, Dallas City skyline at sunset, Texas, USA, dibrova/Shutterstock.com. Pg. 152, DUBLIN, IRELAND—APRIL 1: Grattan Bridge in Dublin, Ireland on the evening of April 1, 2013. This historic bridge spans the River Liffey in Dublin, Ireland, littleny/Shutterstock.com. Pg. 154, Downtown Dubai, zohaib anjum/Shutterstock.com. Pg. 156, Evening view of Pittsburgh from the top of the Duquesne Incline in Mount Washington, Pittsburgh, Pennsylvania, ESB Professional/Shutterstock.com. Pg. 158, Minneapolis. Image of Minneapolis downtown at twilight, Rudy Balasko/Shutterstock.com. Pg. 160, The Saint Joseph Oratory in Montreal, Canada is a National Historic Site of Canada, Click Images/Shutterstock.com. Pg. 162, Independence Hall in Philadelphia, Pennsylvania, f11photo/Shutterstock.com. Pg. 164, Vancouver at night, Dan Breckwoldt/Shutterstock.com. Pg. 166, Salt Lake City, Utah, Andrew Zarivny/Shutterstock.com. Pg. 168, Sunset over Victoria Harbor as viewed atop Victoria Peak, Ronnie Chua/Shutterstock.com. Pg. 170, Skyscrapers by the water in Detroit, Joseph Sohm/ Shutterstock.com. Pg. 172, Tokyo cityscape and Mountain Fuji in Japan, jiratto/Shutterstock.com. Pg. 174, Panoramic aerial view of Gran Via, main shopping street in Madrid, capital of Spain, Europe, Matej Kastelic/Shutterstock.com. Pg. 175, Sunrise View of Portland, Oregon from Pittock Mansion, Josemaria Toscano/ Shutterstock.com. Pg. 178, Panoramic cityscape of Indonesia capital city Jakarta at suny day, Aleksandar Todorovic/Shutterstock.com. Pg. 180, Draper Family, Draper Family. Pg. 181, Socrates,ancient greek philosopher, Anastasios71/Shutterstock.com; Lab, researcher, research, Billion Photos/Shutterstock.com; ST. PETERSBURG, RUSSIA—MARCH 24, 2016: Chief curator of the Literary Museum Ksenia Chudakova holds the presentation for tour operators and media representatives, Lilyana Vynogradova/Shutterstock.com; JULY 10, 2008—BERLIN: the wax figure of Albert Einstein—opening of the waxworks Madame Tussauds, Unter den Linden, Berlin, 360b/Shutterstock.com; prophets of Congonhas / prophets / prophets of Congonhas, lucas nishimoto/Shutterstock.com. Pg. 182, Australia Sidney City CBD close up view over harbour waters at sunset dark sky and reflections of city lights in blurred water, Taras Vyshnya/shutterstock.com. Pg. 184, landscape of kuala lumper, zhu difeng/Shutterstock.com. Pg. 186, Scenic summer evening panorama of the Old Port pier architecture with tall historical sailing ships, yachts and boats and Uspenski Orthodox Cathedral in the Old Town in Helsinki, Finland, Scanrail1/Shutterstock.com. Pg. 188, Nashville, Tennessee, USA downtown skyline on the Cumberland River, ESB Professional/Shutterstock.com. Pg. 190, Scenic summer view of Nyhavn pier with color buildings, ships, yachts and other boats in the Old Town of Copenhagen, Denmark, Scanrail1/Shutterstock.com. Pg. 192, Hamburg- Speicherstadt. Image of Hamburg- Speicherstadt during twilight blue hour, Rudy Balasko/Shutterstock.com. Pg. 194, Moscow, Russia, Red Square view of St. Basil's Cathedral in winter, Reidl/Shutterstock.com. Pg. 196, Close up detail of skyscrapers the business district of Johannesburg—Aerial view of modern buildings of the skyline in South Africa biggest city with South African flag painted on structure walls, View Apart/Shutterstock.com. Pg. 198, Lagos Nigeria, Bill Kret/Shutterstock.com.

About the Author

Michael Moe is the Founder of GSV. He has served as the Chairman, CEO, and Chief Investment Officer of GSV Capital Corp. (Nasdaq: GSVC) since the company's inception in 2011.

Regarded as a preeminent global authority on growth investing, Michael's honors include Institutional Investor's "All American" research team and the Wall Street Journal's "Best on the Street" award. He was recognized by Business Week as "one of the best stock pickers in the country."

Michael conceived of GSV Capital as an opportunity for public investors to invest in premier late-stage, high-growth, venture-backed companies. Driven by the fundamental structural change in the IPO market and in the characteristics of companies going public today, GSV Capital was the first publicly traded security enabling all investors to capture the dramatic growth and value creation taking place in the private marketplace. Noteworthy private invcriestments by GSV Capital include Spotify, Snapchat, Coursera, Facebook, Course Hero, Twitter, Enjoy, Palantir, Lyft, and Dropbox.

Michael has also written extensively about investing in emerging growth-equity markets. In 2006, he authored the business best-seller *Finding the Next Starbucks: How to Identify and Invest in the Hot Stocks of Tomorrow* (Penguin Group), which has gone through three printings in five languages. He also publishes a weekly research newsletter, A2Apple, which focuses on emerging trends in the global growth economy.

Prior to starting GSV, Michael co-founded and served as chairman and CEO of ThinkEquity Partners, an asset management and investment banking firm focused on venture capital, entrepreneurial, and emerging growth companies. Before ThinkEquity, he held positions as Head of Global Growth Research at Merrill Lynch and Head of Growth Research and Strategy at Montgomery Securities.

Michael earned a BA from the University of Minnesota and is a Chartered Financial Analyst (CFA). He is a member of the Board of Directors of OZY Media, Coursera, Curious, GSVlabs, StormWind, and SharesPost. Michael is also a member of the advisory board at Institutional Venture Partners. Michael lives in Silicon Valley with his wife Bonnie and two dogs Bruiser and Rab and ocasionally with his two daughteres Maggie and Caroline.

About the Global Silicon Valley Team

Suzee Han is an analyst at GSV and graduated from Northwestern University with a BA in biological sciences, concentrating in molecular biology. During her time at Northwestern, Suzee lead EPIC, the undergraduate entrepreneurship student group, and created and ran an entrepreneurship development program, a business pitch competition, and an intercollegiate hackathon, among other programs.

Rahul Singireddy is a student at Stanford University majoring in symbolic systems. He is an avid writer with a background in venture capital and product management and is currently working on the newest edition of an entrepreneurship textbook titled Technology Ventures. In his free time, he enjoys traveling, watching the Los Angeles Lakers, and going to Rüfüs Du Sol concerts.

Ryan Demo is a student at Johns Hopkins University studying electrical and computer engineering. Originally and still a graphic designer, he loves bringing to life ideas at the intersection of software engineering and design, often through iOS apps. His favorite pastimes include taking road trips, cooking Italian food, and photographing his friends and family.

Li Jiang is a vice president at GSV and passionate about building start-up communities. He's excited about the possibilities of emerging technologies such as blockchain, mixed reality, machine learning, and microsatellites. He has founded a company, advised entrepreneurs, invested in them, and even made a few documentaries on innovation hot spots around the world. He looks forward to playing ultimate Frisbee with you on Mars.

Nicholas Franco is a vice president at GSV. Prior to GSV, Nicholas served as vice president of business development for ConnectEDU, a venture-backed education technology company, where he led new-market, product innovation, and corporate partnership strategies. Nicholas graduated from Johns Hopkins University with a BA in international relations. He serves on the Network for Teaching Entrepreneurship's (NFTE) National Innovation Committee.